The National Academy Press was created by the National Academy of Sciences to publish the reports issued by the Academy and by the National Academy of Engineering, the Institute of Medicine, and the National Research Council, all operating under the charter granted to the National Academy of Sciences by the Congress of the United States.

Urban Pest Management

A Report Prepared by the

COMMITTEE ON URBAN PEST MANAGEMENT
Environmental Studies Board
Commission on Natural Resources
National Research Council

NATIONAL ACADEMY PRESS
Washington, D.C. 1980

This study was supported by the Office of Research and Development, U.S. Environmental Protection Agency, Contract No. 68-01-2430.

Library of Congress Catalog Card Number 80-83966

International Standard Book Number 0-309-03125-7

Available from

NATIONAL ACADEMY PRESS
2101 Constitution Avenue, N.W.
Washington, D.C. 20418

Printed in the United States of America

iii

Contents

APPENDIXES

List of Illustrations

List of Tables

xii

Foreword

This report is one of a series prepared by the National Research Council for the U.S. Environmental Protection Agency (EPA) in response to Congressional directives to EPA to contract with the National Academy of Sciences to conduct analytical advisory studies. The Congressional directive—which originated in Fiscal Year 1974 in the Subcommittee on Agriculture, Environmental, and Consumer Protection of the House Appropriations Committee—was intended to aid EPA in the formulation of regulations that would reflect not only adequate scientific and technical judgments but also economic and social impacts, and that would overcome the piecemeal approach to pollution control.

The Academy has completed ten studies as part of the National Research Council's Analytical Studies Series. The objective of these studies is to improve the process of environmental management and protection by improving EPA's scientific capabilities in research, monitoring, standard-setting, and enforcement. The studies examined how the Agency applied scientific and technological information in making regulatory decisions. The studies also examined the need for research to remedy deficiencies in the scientific base, in the methods for using scientific information in decision making, and in the manner in which the Agency's needs for scientific advice and information were being met by reliance on external sources and in-house personnel.

Two of the NRC Analytical Studies—Volume II, *Decision Making in the Environmental Protection Agency* (NRC, 1977), and Volume VII, *Pesticide Decision Making* (NRC, 1978)—are of particular significance here because they pointed to past limitations of EPA policies regarding urban pest management and led to the proposal for this study. In particular, the earlier reports showed that EPA had largely limited itself to regulating the use of pesticides in agriculture and had encountered difficulties in considering the distributional effects of environmental regulations and decisions generally. This report is a response to those limitations; it considers urban and inner-city pest management problems, and the role of federal, state, and local agencies in dealing with urban pests and in regulating the use of pesticides in nonagricultural settings. In addition, the study considers the distributional effects of environmental decision making by examining the impact of pest management decisions on the people who live in cities. Particular attention is paid to the protection of the health and safety of inner-city populations, which often suffer from poverty, deteriorated housing, and inadequate municipal services. A special problem of urban pest management, in part a distributional problem as well, is how information on pest management can best be disseminated to urban populations, and what educational efforts are necessary to convey the knowledge and skills of urban pest management essential to protecting public health and improving the quality of life in cities.

REFERENCES

National Research Council (1977) Decision Making in the Environmental Protection Agency. Analytical Studies for the U.S. Environmental Protection Agency, Volume II. Report of the Committee on Environmental Decision Making. Environmental Studies Board, Commission on Natural Resources. Washington, D.C.: National Academy of Sciences.

National Research Council (1978) Pesticide Decision Making. Analytical Studies for the U.S. Environmental Protection Agency, Volume VII. Report of the Committee on Pesticide Decision Making. Board on Agriculture and Renewable Resources, Commission on Natural Resources. Washington, D.C.: National Academy of Sciences.

Preface

The Committee on Urban Pest Management was established by the Environmental Studies Board (ESB) to provide the U.S. Environmental Protection Agency (EPA) with the Board's best judgment of EPA's present and future role in pest control in the nation's urban areas. For the purpose of this study, "urban" means both inner cities and areas that are urban rather than rural or agricultural. The study also responds to ESB's more general concern about the distributional impacts—the costs and benefits and who pays for or enjoys them—of environmental protection. This concern has been expressed in the earlier series of NRC Analytical Studies prepared for EPA—referred to in the Foreword—and in efforts by ESB and its staff to undertake studies of urban environmental problems and their impact on minorities and the economically disadvantaged.

In its request to the National Academy of Sciences, EPA referred to the directions of the Subcommittee on Agriculture, Environmental, and Consumer Protection of the House Appropriations Committee to conduct analytical studies on "the benefits and hazards to humans of agricultural and home use chemicals such as pesticides, herbicides, rodenticides and fertilizers." EPA's request then continued:

In accordance with this general concern and specific mandate, the EPA now requests that the NAS undertake a study of the patterns of pesticide use in the nations's inner cities in order to: (1) determine the extent to which current scientific and technical knowledge are brought to bear on day-to-day pest management practices and (2) develop a clearer picture of the types of health and environmental problems which may be associated with the use of chemical pesticides in the inner city. This study will help the Agency to focus on a poorly understood area which may be directly related to the high incidence of various environmentally induced diseases among our nation's inner city residents.

The EPA request also contained a background statement on the need to ascertain the impacts of urban pesticide use and the potential role of integrated pest management (IPM) techniques in urban areas. This interest in IPM techniques is incorporated in the charge to this Committee:

There is current interest in the use of integrated pest management (IPM) techniques to achieve effective control of pests without sole reliance on chemical pesticides. The Committee will examine various aspects of such an approach including such issues as:
- information needs of urban pesticide users and
- sources and channels of information and technology for urban IPM.

The distributional impacts of pesticide use in urban areas include both public health costs and benefits and economic costs and benefits, and their distribution among city residents. This study may be viewed as a case study of a poorly understood area that has not received much attention in the past, and also as a prototype study of the distributional impact of environmental controls on inner-city residents generally. It has frequently been noted elsewhere that disadvantaged inner-city residents, by reason of multiple exposures in the workplace and the neighborhood, are likely to bear the brunt of the environmental pollution associated with city life, whether such pollution consists of exposure to air and water pollution or to pest-borne diseases associated with the crowded conditions of inner-city life.

Given the broad task delegated to it, the Committee had first to resolve a number of difficulties in defining the scope of its work. In resolving them, however, a number of insights were gained that we believe inform and lend substance to the report. While there was general agreement that the management of pests in an agricultural or rural setting differs significantly from the management of pests in urban areas, it was necessary to define the term "urban." The inner city is clearly urban, but urban areas differ significantly throughout the United States. The term "inner city" finds its clearest expression in the older cities of the Northeast, such as Boston, Philadelphia, and New York, but it loses much of that clarity when applied to southwestern and western cities, such as Los Angeles, Phoenix, and Sacramento. In many cities, moreover, the line between the inner city and the suburbs is hard to draw. Managing pests in suburban gardens and lawns is clearly different from managing pests in an agricultural setting involving plant monoculture; it also lacks the principal concern about public health that characterizes pest management in the inner city.

The problem of drawing lines also was complicated by the diverse interests of members of the Committee. For some members the primary issue was how best to deal with pests in the inner city for the protection of public health. Other members saw a more significant task in the study of pest management for economic and aesthetic purposes in the general urban environment. It was clear that in order to undertake either task meaningfully the nature of the pest problem—whether in the inner city or in the urban environment generally—had to be examined, because part of the difficulty is in determining what is to be managed. Stated simply, it is important to know which is the bigger problem in the cities: the presence of rodent and insect pests or the effects of the toxic substances used to dispose of them. This question is less often asked in the agricultural setting because pest management is accepted as an integral and necessary part of the production of food and fiber and of animal husbandry.

In resolving definitional issues of the scope of the study, the Committee decided to deal both with pest management issues in urban areas generally, encompassing economic and aesthetic problems, and with pest control in the inner city, emphasizing the protection of public health.

The problems of urban pest management differ from other pest management problems because city life—particularly life in densely populated inner-city environments—brings people and pests into close proximity. The interaction of people and pests, and the psychological attitudes of people toward pests, thus become important subjects for study. Psychological and social attitudes are likely to affect behavior and the readiness to change behavior as new information about urban pest management becomes available. Thus, study of the psychological response to pests is crucial in planning new management strategies that de-emphasize immediate recourse to toxic chemical pesticides.

The general purposes of this study, as formulated by the Committee, are: (1) to examine and evaluate the problems created by a variety of pests in various urban areas; (2) to study and evaluate current efforts to manage and control pests in urban areas by examining the role of federal, state, and local agencies, and private, nongovernmental activities; (3) to examine popular attitudes toward pests in urban areas and to examine the effectiveness of efforts to convey pest management knowledge and skills to urban residents; (4) to determine the importance of urban pest management in relation to other urban environmental problems; and (5) to offer recommendations for the improved management of pests in urban areas to enhance human health and the quality of urban life.

In examining these issues the Committee considered the feasibility of transferring technology from the highly developed field of agricultural pest management to the highly specialized problems of urban pest management. Technology transfer is a complicated problem, however, because most urban pest control activity is carried out by individual householders in large and heterogeneous populations. The Committee also sought to deal with the adequacy of legal and regulatory authorizations, in particular with the question of whether current laws and other regulatory controls and their present interpretations are best designed to advance urban pest management or whether changes are called for.

This report has five chapters and an Executive Summary that presents the Committee's findings and recommendations. Chapter 1, an overview of the study, includes a discussion of the purpose of the study, brief analytical discussions of critical issues in urban pest management, an historical perspective on urban pest management, a discussion of the relationship of urban pest management to other environmental disciplines, and a description of the scope of the study.[1] Chapter 2 is a summary description of the impact of major urban pests on the human population in inner-city and suburban settings—the plant and animal species that affect human health, property and structures, and aesthetic and recreational values. The chapter also examines the problems of perceptions and attitudes about pests. Chapter 3 examines a variety of strategies for managing urban pests, including the use of chemical pesticides and the use of integrated pest management (IPM) strategies. Chapter 4 reviews the economic and distributional aspects of these strategies. Chapter 5 focuses on decision making and the role of federal, state, and local governments in managing urban pests. The chapter also addresses the problems of information transfer and community participation in urban pest management.

The Committee wishes to thank those who contributed to the study, including government and international organizations; environmental and public interest organizations; professional, scientific, and trade organizations; consultant, business, and industry organizations; and individuals. Their names appear in Appendix A.

The Committee also extends special thanks to those who served as consultants or subcontractors to the study. Their contributions and advice were also useful to the Committee in preparing this report. They include:

DONALD COCHRAN, Virginia Polytechnic Institute
TOM DAVENPORT, University of Maryland
RODNEY DE GROOT, USDA Forest Products Laboratory,
    Madison, Wisconsin
LESTER EHLER, University of California, Davis
CLYDE ELMORE, University of California, Davis

RICHARD V. FARACE, Michigan State University
CLARENCE FAULKNER, U.S. Department of the Interior
WILLIAM FEIST, USDA Forest Products Laboratory, Madison, Wisconsin
MARY LOUISE FLINT, University of California, Berkeley
JACK FRASER, University of California, Berkeley
JOE GOOD, U.S. Department of Agriculture
NANCY GOREN, University of California, Berkeley
ROGER GROTHAUSE, Armed Forces Pest Management Board
WILLIAM HELMS, U.S. Department of Agriculture
LENNEAL HENDERSON, Howard University
FREDA HORAY, RvR Consultants, Shawnee Mission, Kansas
MARK ISSACS, U.S. Department of Housing and Urban Development
KARL KAPPUS, U.S. Public Health Service, Center for Disease Control
STANISLAV KASL, Yale University
BENJAMIN KEH, California State Department of Health Services
CARL KOEHLER, University of California, Berkeley
E. PHILLIP LEVEEN, Public Interest Economics-West Corporation
VERNARD LEWIS, University of California, Berkeley
HARVEY SAPOLSKY, Massachusetts Institute of Technology
INGRID SMITH, University of Maryland
STEVE SPAULDING, Bowling Green State University
EUGENE THOMPSON, Armed Forces Pest Management Board
SANFORD WEINER, Massachusetts Institute of Technology
GAYLE WORF, University of Wisconsin, Madison.

A list of working papers made available to the Committee is given in Appendix B.

We are indebted, too, to the EPA staff, particularly David Andrews, Bill Dickinson, Mike Dover, Judy Heckman, Ken Hood, Mike Moore, Jim Wilson, Charles Reese, and Doug Sutherland, who assisted the Committee in obtaining information essential to its task.

On behalf of the Committee, I would like to thank Lawrence Wallace and Elizabeth Panos of the National Research Council for their special contributions in preparing this report and for their competent, unstinting, ever-cheerful, and professional assistance to the Committee, and especially to its Chairman. Other staff members who provided assistance include Judith Cummings, Estelle Miller, Christina Olson, and Robert Rooney, and especially Raphael Kasper, who helped the Committee and its Chairman over the institutional hurdles involved in preparing this report. Dorothy and Philip Sawicki provided thoughtful and painstaking editorial assistance.

Finally, I wish to express my personal thanks to each member of the

Committee for his or her devoted efforts toward successful completion of the study, for outstanding professional competence, and for patience and cooperation in preparing our recommendations.

FRANK P. GRAD, *Chairman*
Committee on Urban Pest Management

NOTE

1. One Committee member, Gordon Frankie, disagrees with the relative emphasis given to inner-city pest problems and suburban pest problems, especially in the Preface and Chapter 1, and writes:

During the short official period (about 8 months) that the Urban Pest Management Committee was operative, it had to deal with a great number of problems in gathering information. The most difficult problem was the unrealistic time frame within which the study was to be completed. When the report was finally assembled there was little time left for the Committee to effectively analyze, integrate, and critique the various pieces. Thus, committee members ceased to interact as a group in an effective way during the final phase of the report preparation. Because of this problem, I feel that a significant piece of the introductory material (*Executive Summary through the definition of Urban in Chapter 1*) could have been more effectively reviewed by the Committee. In the following paragraphs I would like to comment on what I believe are loosely worded generalizations that may lead to misunderstandings on the part of the uninformed reader.

To begin, I was never satisfied with the attempt to define the term, Inner City. In fact, it is stated in the Introduction that the inner city concept probably applies mostly to the northeastern United States. Despite this problem, the Introduction proceeds ahead giving the impression that the Committee had reached agreement on some characterization of this term, which it did not. The result then, is a rather freely used term, Inner City, throughout much of the Introduction.

The use of this term can perhaps be justified conceptually when appropriate qualifications are applied. However, it is clear that little qualification was offered in the introductory material. Because of this situation, I would like to caution the reader to a number of difficulties, the most serious of which concerns the classification of particular pest groups according to certain sections of the city, i.e., inner versus other sections (including suburbs) of the city. It is stated that inner-city residents are mostly bothered by public health problems and that non-inner-city residents are bothered primarily by economic and aesthetic problems. I maintain that this categorization developed in the minds of certain committee members during the various meetings and somehow became entrenched in their respective writing assignments. Results of actual survey work (limited) on the various pests, which are presented in Chapter 2, *did not* indicate that pest groups could be meaningfully sorted according to particular sections of the city. Rather, the findings of this work suggest that carefully planned surveys are needed to determine the distribution and frequency of public health and nonpublic health pests within and between urban areas.

The information that was gathered indicated that most public health problems can be found in a variety of locations within cities. In this regard, no studies could be found that provided information on substantial differences in severity of public health problems within cities. Further, economic and aesthetic problems can be found in all parts of a city, and there were no data to indicate that these problems are mostly found in outer city areas. Finally, the Introduction leaves the impression that inner-city people are more exposed to pesticides. However, as clearly indicated in a later chapter, evidence to support this generalization is virtually nonexistent. Unfortunately, the Introduction did not explore the possibility that pesticide usage may be extensive in other affluent areas of cities, which in turn would pose a special kind of public health problem.

Overall then, the above problems represent the fitting of impressions into a vaguely conceived mold of the inner versus the outer city. The information provided in this particular introductory section does not necessarily reflect the state of the knowledge, nor does it reflect what appears later in the report.

I would also like to comment on one other aspect of the report. In the Preface it is stated that the scope of the study involved pests of public health and non-public health importance. One therefore is led to believe that there is a balance between these two major pest groupings (early on, the Committee agreed that the report should be balanced). However, as one proceeds through the introductory material, it becomes clear that concern over public health problems dominates. In fact, in two instances in the Introduction it is stated that public health problems deserve priority (i.e., over non-public health problems). This obvious bias does not reflect a consensus of opinion. The public health bias in this report is manifested in another way. During the formative stages, the Committee decided that only certain major groups of pests would be examined. These pests included plant pathogens, weeds, and wood-destroying organisms. Yet, after the Committee held its last meeting, information on these latter pests was relegated to working papers, rather than being included as part of the main report. Members were later informed that the action was taken because of a lack of funds to process the information into the report. However, it should be mentioned that no public health information was sacrificed at any stage in the development of the report.

Executive Summary

This investigation of urban pest management has a dual purpose—first, to study urban pests and the risks they pose to human health and to environmental, economic, and aesthetic interests, and, second, to study efforts to manage and control such pests and to examine the risks, costs, and benefits that result from the various methods and techniques that have been or can be applied in that effort. The Committee faced a number of problems in defining the scope of the inquiry. What precisely is the "urban" environment in which pests are to be managed or controlled, and why is the management of pests in an urban environment significantly different from the management of pests in an agricultural setting? An examination of these issues, in turn, led to the question of the aims and purposes of urban pest management.

The urban settings covered in this report include not only the densely populated inner city, deteriorated or not, but also the outer and suburban rim of the city or metropolitan area. There is an essential difference between pest problems in different parts of a city. Pest problems in the outer rim more frequently involve damage of an economic or aesthetic nature—e.g., to structures, to lawns, to ornamental plants—while inner-city pest problems more frequently involve major threats to public health. Thus, the primary purpose of urban pest management in the inner city is protection of human health and life, while in the outer part of the city it is avoidance of economic and aesthetic damage. Both purposes involve protection of aspects of the quality of life and are essentially different from the purpose of agricultural pest management, where the purpose is primarily economic, being integrally connected with the growing of crops. As a result, the pest controller in agriculture is likely to be a professional or semi-professional—a circumstance that is recognized in federal pesticide legislation, which seeks, in the first instance, to protect persons exposed to pesticides because of their occupation.

The presence of pests in the inner city is different because the rats, cockroaches, fleas, lice, flies, and other pests there are part of the general living conditions in the inner city. Certain pests are closely associated with densely packed populations, housing decay, inadequate management of wastes, and poor habits of sanitation. They are symptomatic of, if not causally connected with, unemployment, racial discrimination, and a general feeling of hopelessness in some inner-city communities, a hopelessness that makes it difficult to take action to remedy these adverse conditions. Pest management in the inner city must therefore deal not only with the environmental problems of pests and pesticides, but it must also concern itself with social and economic conditions in their entirety.

The heterogeneity of city populations greatly affects urban pest management. Information and education on pest management in agriculture is addressed to people who share common interests and similar values, but city populations are more diversified and, in many instances, more at risk, especially from the use of pesticides. Urban residents are seldom trained to deal with hazardous pesticides in a knowledgeable, professional way. Inner-city populations, in consequence, are more exposed, it seems, to excessive pesticide exposure, just as they are more exposed to pests that are carriers or vectors of disease. Indeed, the exposure of the urban poor to pests and pesticides seems similar to their greater exposure to environmental pollution generally.

One of the major difficulties in urban pest management is its relative newness. Although there are many studies of pest management in an agricultural setting, urban pest management has only recently been recognized as a separate and specialized field. Many studies of pests and of pesticide use and of the dangers and consequences of such use do not segregate city or metropolitan data, and the literature devotes little attention to the public health hazards posed by urban pests, perhaps because the virtual disappearance of many of the communicable diseases common a few years ago has lulled the public and policy makers into a false sense of security.

MAJOR URBAN PESTS

There are two general categories of urban pest species—those that adversely affect people's health and well-being and those that adversely affect individuals' economic and aesthetic values. Chapter 2 surveys major urban pest species, with substantial attention to the distribution, habitat, and significance of selected arthropods of medical and veterinary importance. Also surveyed are pests that affect economic and aesthetic values, particularly major indoor and outdoor urban arthropod species. (Working papers made available to the Committee provided detailed information on a variety of nonarthropod pests, such as plant pathogens, weeds, and wood-destroying pests other than termites. See Appendix B.)

Although both types of pests are important, controlling pests that are a threat to public health may be a matter of greater urgency. The Committee's concern with public health pests arises largely from the possibility of significant outbreaks of pest-borne diseases, such as plague, dengue, yellow fever, encephalomyelitis, and malaria, as well as outbreaks of variant, heretofore unrecognized, or relatively new diseases, such as rickettsial pox, babesiosis, and Lyme arthritis. The overcrowded and filthy conditions in many inner cities provide a habitat conducive to the propagation of a number of pests that are important disease vectors. In addition, suburban areas—with their increasing proximity to woodlands and other dense vegetation—provide a habitat for a number of arthropod and vertebrate pest species that can cause human injury and illness. When humans live in proximity to animals that harbor infectious agents that may also infect humans, the hazard that those agents may be transmitted to humans must be recognized.

Vertebrate and arthropod pests cause several hundred deaths in the United States each year. The actual number is not known, however, because the diseases in question are often not recognized or diagnosed. In addition, many other people suffer pain, annoyance, disfigurement, emotional distress, or disabling conditions as the result of bites, stings, or physical reactions to pests and their excreta (see Chapter 2).

Urban pests have different impacts on different segments of the urban population. Suburban dwellers are more likely to encounter such diseases as Rocky Mountain spotted fever, tularemia, sylvatic plague, and rabies; such venomous arthropods as bees, hornets, wasps, chiggers, and fire ants; and such mosquito-borne diseases as encephalomyelitis. Commensal pests, such as rats, mice, flies, cockroaches, and a number of human ectoparasites are more commonly found in densely populated inner cities where sanitation is poor and where buildings and other structures provide suitable environments. For a variety of reasons, a majority of the diseases transmitted by commensal pests and human ectoparasites are encountered most by poor people and by those who inhabit or work in the inner cities.

Although there are clear regional differences in the distribution and relative importance of many pests, detailed assessments of their impact on public health and of their effect on economic and aesthetic values cannot be made because of a lack of data on the occurrence and prevalence of urban pests nationwide (see Chapter 2). The problem stems largely from insufficient attention to urban pest problems in the past, caused in part by the fact that no federal agency has had direct responsibility for the management of pests in urban areas. The decrease, and in some cases the virtual disappearance, of pest-borne diseases has led to diminished public and governmental awareness of the threat still posed by these diseases. Large reservoirs of major pest-borne diseases exist in the United States and in many other parts of the world, and potential vectors of these diseases live in close association with large numbers of susceptible people. Air transportation increases the risk that the agents that cause life-endangering diseases, including some not hitherto present in this country, may be introduced into our population.

• *We recommend that health agencies at every level of government emphasize their traditional role in the prevention of disease, and that they develop and engage in innovative programs to control public health pests.*

Since 1965, when health departments became increasingly involved in the provision of medicare and medicaid treatment services, the traditional functions of sanitary control and environmental health have largely been subordinated among health department priorities. The need to control pests calls for renewed attention to this traditional area of concern. Innovative programs in the field might include measures to improve the urban environment by eliminating conditions that support vectors of disease and give rise to the use of toxic chemical substances in the environment, and by stimulating or engaging in programs of environmental modification and integrated pest management.

• *We recommend that the reporting and surveillance system developed under the guidance of the Center for Disease Control of the U.S. Public Health Service in collaboration with state and local health departments be supported and strengthened. We further recommend that the U.S. Public Health Service be provided with the funds and administrative authority necessary to carry out adequate surveillance and inspection of travelers, goods, and other potential carriers of disease into this country and to provide information on the incidence and prevalence of pest-related diseases, which in turn will provide an adequate basis for the evaluation of methods of control and prevention.*

• *We recommend that the U.S. Department of Health, Education, and Welfare*[1] *expand its Bureau of Community Environmental Management to protect people from pests that attack them directly or otherwise threaten their health and well-being. We further recommend that the Bureau's programs be developed in close collaboration with other national, state, and local government agencies that have an important interest in the problems posed by these pests.*

This effort should include collaboration with the U.S. Department of Housing and Urban Development on the construction and maintenance of pest-resistant urban structures and the development of programs for the control of pests in public buildings, parks, and streets; with the U.S. Department of Labor to provide protection from pests in working places; with the U.S. Department of Education in developing programs for the control of such pests as scabies mites and head lice, which infest school children and adolescents, and in creating educational programs for schools and community groups that focus on pests and pest management; with the U.S. Department of Agriculture in developing programs for the control of pests that attack agricultural products, plants, animals, and buildings; and with the U.S. Environmental Protection Agency in its programs for regulating pesticides.

The needs and desires of community groups should be ascertained before such programs are initiated, and the programs should be developed with the assistance of scientists and professionals in pest management who are able to make appropriate use of methods that are likely to achieve the long-term goals of reducing human exposure to the adverse effects of pests while at the same time reducing the use of hazardous chemicals and reducing the dangers to the environment and other species of animals or plants.

• *To assist in the development of these programs, we recommend increased research efforts to study the ecology and control of pests that endanger people; increased federal support to state and local agencies and community groups to develop and maintain pest control programs; and the inclusion of pest control considerations in existing federal programs to assist poor persons and families to live in decent housing free from health-threatening pests and to maintain sanitary conditions to achieve and sustain decent standards of health.*

• *We recommend studies of a number of social aspects of urban pest management, including*

1. *The relationship between social-psychological factors (e.g., demographics, habits, group norms) and perception, incidence, and management of pests;*

2. *The development of effective intervention strategies (e.g., education, incentives, regulations) sensitive to social-psychological factors;*

3. *Evaluation of the effectiveness of such intervention strategies (outcome research) and examination of the dynamic interaction among intervention efforts, behavior, attitudes, and incidence (process research);* and

4. *Consumer and administrative decision making related to urban pest management.*

Attitudes and behavior toward pests have a direct bearing on the selection of control methods and the willingness to adopt alternative pest management strategies. Despite recent scientific interest in generalized attitudes and behavior toward environmental problems, urban dwellers' perceptions of pests, pesticide use, and how they are affected by pests and pest management practices are relatively unknown (see Chapter 2).

MANAGEMENT OF URBAN PESTS

Information on the kinds and amounts of chemical pesticides applied in urban areas by various user groups is essential to the interpretation of epidemiological and monitoring data and to the evaluation and possible improvement of current pest management programs. The Committee's review of chemical management methods is based on an examination of data on nonfarm pesticide use at the national, state, and local levels (see Chapter 3). The principal aim was to examine patterns of pesticide use in order to develop a clearer picture of the types of health and environmental problems associated with the use of chemical pesticides in inner cities.

• *We recommend a comprehensive nationwide study of pesticide use in urban areas. The study should focus on*

1. *The amounts and kinds of pesticides used, both inside the home and in adjacent yards, gardens, and other open areas;*

2. *The extent of professional and individual household application of pesticides;*

3. *Human exposure to pesticides;* and

4. *Morbidity and mortality from urban pesticide use.*

Little attention has been paid to the quantitative aspects of pesticide use in urban areas or to the fate and disposition of pesticides applied for nonagricultural purposes. Of the studies reviewed for this report, none provide data on the chemical active ingredients or quantities of pesticides used. Nor do any of them cover the four major user groups involved (households, pest control operators, public agencies, and commercial and industrial users), and most were designed and conducted primarily for purposes other than the collection of information on pesticide use in urban areas. Thus the data assembled to date are inadequate to allow quantitative analysis of pesticide use in urban areas in the United States.

Concern about the effects of pesticides on the health of urban residents is based primarily on documented harmful effects in agricultural settings. The principal health concerns regarding urban pesticide use include determining the extent of human poisonings by pesticides and the effects of chronic or incidental exposures. Although limited, the data clearly demonstrate the possibility of sizeable and multiple exposures to chemicals used in urban pest management programs (see Chapter 3).

• *We recommend a detailed examination of a variety of alternative approaches to urban pest problems.*

The Committee's review of alternative pest management techniques was prompted by the need to reduce reliance on chemicals to control urban pests (see Chapter 3). Although the application of integrated pest management (IPM) strategies has not yet been well explored, such strategies may reduce the exposure of urban populations to chemical pesticides. A number of alternative methods—biological control, host-plant resistance, insect growth regulators, genetic manipulation, habitat modification, and a variety of cultural, mechanical, and physical controls—have been examined. IPM appears to be a promising direction, but it must be noted that IPM may not be appropriate where the objective is total eradication of pests because of their effect on public health. Although an in-depth look at alternative strategies was beyond the scope of this study, integrated efforts should be encouraged whenever they seem feasible. More information on both demonstration projects and operational programs is needed.

• *We recommend studies of urban pest organisms to develop basic biological, behavioral, and ecological information that can be applied in integrated pest management strategies.*

ECONOMICS OF URBAN PEST MANAGEMENT

The Committee's economic analysis deals primarily with theoretical aspects of the social efficiency of current urban pest management practices (see Chapter 4). There are strong reasons to believe that the private market alone cannot produce socially efficient solutions, and that society may incur substantial unnecessary costs if the private market is allowed to operate without governmental restraint. There is particular concern about the distributional impact of current pest management strategies on poor inner-city residents. Two major issues arise in any attempt to assess the economics of urban pest management—whether the level of resource allocation is appropriate, and whether current management practices are cost-effective.

Because of the lack of data, the Committee's analysis is largely conceptual. A number of research priorities for remedying the situation are identified.

• *To facilitate a more detailed analysis of the economic aspects of urban pest management, we recommend studies on*:

1. *The costs of urban pest management and their distribution among government and private sources*;
2. *The benefits of urban pest management and assessment of the comparative benefits of different control methods, including consideration of the cost-effectiveness of the modifications in legislation on housing and waste control that would be required*;
3. *The structuring of incentives for better urban pest management*;
4. *The results of urban pest management programs, including evaluation of their relative success or lack of success*; and
5. *Health and economic damages caused by urban pests.*

URBAN PEST MANAGEMENT DECISION MAKING

Examination of federal, state, and local statutes and regulations led to the conclusion that urban pest management has not yet found a home in the law (see Chapter 5). Furthermore, Congress and the Federal Government have not developed a policy on urban pest management. Although there is considerable interest in the regulation and management of pesticides, there is literally not a single mention of urban pest management, with the possible exception of rodent control, in the federal statutes. Pest management in the agricultural setting, on the other hand, is a frequent subject of federal legislation, with detailed research and management authority vested in the U.S. Department of Agriculture and in the U.S. Department of the Interior. Although legislation authorizing research on agricultural pests is often of benefit to urban pest management as well, that is not its primary purpose. Other federal legislation, particularly legislation on solid-waste management, and its impact on state and local activities also is relevant, but, again, it is not expressly directed at urban pest management. Needs for rodent control are reflected in the legislation authorizing the U.S. Department of Health, Education, and Welfare's rat control program and in laws on urban renewal administered by the U.S. Department of Housing and Urban Development, but this is the extent of explicit federal involvement. The U.S. Food and Drug Administration, the U.S. Department of the Interior, the U.S. Department of Defense, and the General Services Administration carry out pest control and pest management in their urban facilities, but they have no regulatory authority and their activities are neither systematic nor uniform.

Although state governments also regulate pesticides, subject to the controlling authority of federal law, there is little in state law pertaining directly to urban pest management. The 1978 amendments to the federal pesticide legislation (FIFRA—Federal Insecticide, Fungicide, and Rodenticide Act, as amended) gave the states primary responsibility for enforcing the law against pesticide use violations, but this authority (as well as state cooperative extension service training of pesticide applicators) is primarily addressed to agricultural pesticide uses.

State authority to control mosquitoes is sometimes delegated to local agencies or to mosquito control districts, and control of rodents and other insects is commonly delegated to local health agencies or to local agencies that enforce housing standards. Sometimes the authority is implied in legislation that authorizes the abatement of public health nuisances.

There are also state laws on solid-waste management that impose requirements on local governments relating to the collection and disposal of solid wastes and to the regulation of sanitary landfills pursuant to federal guidelines and regulations under the Resource Recovery and Conservation Act of 1976. These requirements commonly include provisions for pest control, particularly of rodents. Here again, however, urban pest management is not an identifiable state concern.

It is at the local level of government that most urban pest management activities are implemented and regulated. Normally, pest control activities are carried out under local health or housing codes, generally under state

enabling legislation. Housing codes usually contain rat control requirements. The nuisance abatement powers of local government are also relied
on for pest control enforcement, and some local health agencies initiate
and support educational programs on pest management and related
subjects.

Local pest management activities are frequently directed toward high-
risk sites, such as public buildings, transportation facilities, food establishments, industrial sites, and, most importantly, multifamily housing, in the
attempt to control such target species as rats, mosquitoes, flies, lice, ticks,
mites, bedbugs, fleas, cockroaches, and birds. In addition, many local
governments enforce state legislation relating to structural pests.

Although some local governments assert that they apply integrated pest
management techniques, there is nothing in local legislation that requires
them to do so. Most local laws require the householder in a single-family
dwelling and the owner of a multiple dwelling to be responsible for pest
control on the property, but there is no control over the method used.

Thus, while there is considerable recognition of urban pest problems in
local law, those laws are variegated in content and quality, and fail to
address urban pest management problems generally. The picture is one of
fragmentation, with specific programs addressed to different kinds of pests
and with authority widely dispersed among different agencies. To some
extent the fragmentation is the result of state legislation that delegates and
authorizes the exercise of pest management and vector control functions in
a haphazard and unorganized fashion.

• *We recommend that the subject of urban pest management be
adequately reflected in federal and state legislation, and that such
legislation address the needs for more research and improved interagency
coordination.*

At present, several federal departments and agencies have responsibilities relevant to urban pest management. EPA has general authority to
regulate the manufacture and use of pesticides; HUD is concerned with
the improvement of deteriorated housing that has great relevance to pest
control in urban areas; the Department of Agriculture has broad research
and educational responsibilities; and HEW has a long history of concern
for vector control and for the prevention and control of human disease. It
is clear that any concerted effort to deal with urban pests and urban pest
management problems will have to involve contributions from each of
them. But one of the most significant problems is that no one federal
agency has had specific overall responsibility for urban pest management.

• *We recommend, therefore, that one department or agency be designated to coordinate the urban pest management initiatives of the departments and agencies noted above, and to be responsible for administering the grant-in-aid program recommended below.*

• *To enhance federal-state interaction in urban pest management, we recommend that the Federal Government establish a grant-in-aid program under the direction of a designated federal department or agency to provide funding to urban areas for the management of pests, and that such a program contain provisions requiring states and municipalities to have and enforce adequate laws and regulations for the management of pests as a condition of obtaining federal funding.*

To the extent appropriate, existing programs of environmental modification should be combined with the grant-in-aid program. The present rodent control program and the earlier "workable program" requirements of federal housing law could provide the pattern for grant-in-aid legislation, as could federal law requiring state solid-waste plans as a condition of federal funding assistance for solid-waste management facilities.

• *We recommend that uniform federal guidelines for urban pest management be established and promulgated by appropriate regulation, following coordinated efforts and review by federal agencies and departments.*

• *We recommend that the states enact legislation to create comprehensive pest management programs that take into account the effects of health, housing, environmental, and waste disposal controls on urban pest management.*

• *We recommend that major initiatives be undertaken to develop appropriate research and educational courses in urban pest management and to support public education programs on urban pests and urban pest management, and on related public health issues generally.*

Examination of governmental initiatives in urban pest management reveals a need for more knowledge about effective means of communicating pest management information (see Chapter 5). Urban pest management relies in part for its development on effective communication of information and techniques to virtually every segment of the population.

This need has not been adequately met at any level. The universities have produced little useful scholarship on the subject, and attempts to educate the general population about urban pest problems have been inadequate.

- *We recommend that officials of the departments and agencies involved in urban pest management ensure that the principles set forth in this report are used to the fullest extent in all publicly funded pest prevention or control activities.*

- *We recommend that the urban pests named in this report be given full consideration as pests to be brought under control to improve the quality of life in urban areas.*

Section 28 of FIFRA, as amended, requires the EPA Administrator, in coordination with the Secretary of Agriculture, to identify those pests that must be brought under control. Such coordination should reflect more concern with problems posed by urban pests, particularly those of public health significance.

It should be noted that since many of the Committee's recommendations, if implemented, will require some level of federal support, consideration of the costs of implementing the recommendations is important, especially in the present climate of fiscal restraint.

NOTE

1. After the drafting of this report the name of the Department of Health, Education, and Welfare was changed to the Department of Health and Human Services. Those recommendations and discussions that refer to the Department of Health, Education, and Welfare apply as well to the new Department of Health and Human Services.

1

Introduction

PURPOSE OF THE STUDY

The chief purpose of this study is to highlight and examine an environmental, economic, and public health problem that has received too little attention to date, namely, the widespread existence of pests in the urban environment.

A variety of other environmental problems—air and water pollution, disposal of solid wastes, and excessive noise—have been major targets for environmental control at every level of government, and their nature as predominantly urban problems has long been recognized. Part of the reason for governmental efforts to abate them has been the clear evidence of their negative impact on the health, comfort, and quality of life, and on the economic and aesthetic interests of the city dweller. But pests in the urban environment have received far less attention, particularly from the Federal Government, even though their impact on city dwellers is similar in many respects to that of the more readily recognized urban pollutants. Pests, too, have an impact on the health, comfort, and quality of life, and on the economic and aesthetic interests of the city dweller.

Public health officials, to be sure, have long had an interest in rodent and mosquito control, and school officials in eliminating the problems of bedbugs and lice. Housing authorities and financial institutions have long been concerned with protecting homes against termites and other wood-destroying pests. But pest control as an integrated, systematic field of knowledge and regulatory control has chiefly been considered a rural and agricultural field of expertise, primarily concerned with the economics of animal husbandry and the growing of food and fiber. The more recent awareness of urban pest management problems, reflected in the charge to this Committee, is in part due to increasing concern about human exposure to toxic substances. That concern is evidenced by such recent legislation as the Toxic Substances Control Act of 1976 (Public Law 94-469, 90 Stat. 2003, 15 U.S.C. Sec. 2601). Although human exposure to pesticides has been a concern of federal pesticide control laws since 1910,[1] the concern is primarily with occupational exposure.

Interest in urban pest management is also part of a growing concern for the development of effective environmental controls in cities. In an increasingly urban society, expenditures for environmental controls must be justified largely in terms of their effectiveness in making the urban environment more livable. The focus on industrial pollutants such as asbestos and benzene reflects an essentially urban emphasis in air pollution control, as does the emphasis on automotive pollution. In the recent past there has been increased interest in controlling the emission of lead, which has had a major impact on the children of poor persons who live in the worst parts of our cities.[2] The construction of public waste water treatment plants to serve major urban areas has been a multibillion dollar federal priority since the enactment of the Federal Water Pollution Control Act in 1972 (Public Law 92-500, Title II, Sec. 201, 33 U.S.C., Sec. 1281 et seq.). More recently, major attention in water pollution control has been given to toxic pollutants, primarily industrial and urban in origin, that affect the safety and purity of drinking water supplies,[3] another problem substantially urban in character.

Problems of solid-waste disposal—problems that Congress has recognized as a direct outgrowth of urbanization and industrialization—have also occasioned national concern, not only because solid waste represents a possible misuse of reusable resources and a potential waste of scarce energy, but also because some urban areas, particularly major cities, are about to run out of sites for sanitary landfills (Resource Conservation and Recovery Act of 1976, Public Law 94-580, Sec. 1002, Congressional Findings, 42 U.S.C., Sec. 300h-l). The Federal Government has not only undertaken a national program to develop sound methods of waste management, but it has also begun a major effort to deal with hazardous wastes (Resource Conservation and Recovery Act of 1976, Public Law 94-580, Sec. 3001 et seq., 42 U.S.C., Sec. 6291 et seq.). The management of waste in cities has relevance to the hygiene of housing and to the management of many urban pests that find harborage and food in improperly managed wastes, and become carriers of pathogens that cause human disease.[4]

Housing and the hygiene of housing have sometimes been artificially separated from other environmental concerns in the structure of government programs, even though housing has the most abiding environmental impact on human life. The efforts of society and government to improve housing, particularly in urban areas, constitute one of the oldest "environmental" programs, and although it was begun in the late nineteenth century, the program is still far from complete. A substantial number of people are still far from enjoying the decent home for every American that was reemphasized and set as the target for the housing effort by the Douglas Commission in 1969.[5] The poorest of our population still live in dilapidated and unsanitary housing surrounded by accumulations of garbage, filth, and squalor that encourage the propagation of rodents and other pests. As recent studies[6] have shown, the deprived inner city subjects the poor to a variety of environmental insults, including excessive amounts of air and water pollution, pest-borne diseases, rat bites, and toxic substances (often resulting from efforts to control pests). These conditions are often accompanied by inadequate public services, particularly for the collection and removal of garbage and other wastes. Thus, pest management in the inner city must be viewed in the context of many other problems.

Yet it is clear that dilapidated and overcrowded inner cities do not constitute all, or even a major part, of urban areas. Many old parts of cities do not share the inner-city blight just described. Thus, pest problems, like all other environmental problems, have different characteristics in different settings. Urban pest management has different meaning, therefore, in different parts of urban America. There are, indeed, pest management problems in city suburbs and in other parts of cities with lesser population densities, where single residences with open areas, gardens, or back yards are prevalent. The economic and aesthetic interests in protecting urban lawns, gardens, ornamental trees and shrubs, and in safeguarding golf courses and other open areas is real and will be dealt with in this report along with inner-city pest problems.

In many urban environments the concept of "pest management" is more appropriate than "pest control." Pest management is addressed to maintaining low and balanced populations of biota that subsist on lawns and in urban gardens and open areas. However, in inner-city areas, which usually have little open space, pest control or pest eradication is probably the more appropriate term. It is difficult to view a balanced population (low or high) of rats, cockroaches, fleas, and lice as an aim of inner-city pest control. Eradication of these organisms, however, is rarely achieved and is virtually impossible to sustain.

Modifying the behavior of urban dwellers is a necessary concomitant of all urban pest management efforts, but behavioral control efforts face special problems in the inner city, where poor housing conditions, high population density, and inadequate municipal services often make behavioral change particularly difficult. Economic deprivation and an ensuing sense of hopelessness make it unlikely that instruction in the management of garbage or the latest integrated pest management (IPM) methods will have significant effects. Although efforts to involve inner-city populations in pest control may eventually be effective, they will probably have a better chance of success as part of a broader effort to improve living conditions.[7]

Integrated pest management in the inner city largely consists of environmental modification "to a point where [the environment] is no longer suitable for the breeding or development of insect vectors or of rodent reservoirs of disease, e.g., improved drainage for the control of mosquitoes, and proper refuse disposal to prevent the breeding of flies and rodents" (World Health Organization 1972). Thus, IPM and vector control in the inner city rely on similar mechanisms and may in effect be synonymous.

Another purpose of this study is to examine the costs and benefits of various approaches to urban pest management. Urban pest management, both for the protection of public health and for the maintenance of economic and aesthetic values, is often costly in economic terms, and many management and control strategies may themselves pose risks to health and involve other costs. The question of the appropriate distribution of costs in the light of particular benefits must therefore be examined, as must the question of priorities of incurring the costs for particular purposes and aspects of urban pest management. Thus pest management in the inner city is subject to different cost considerations than is pest management in areas where public health problems are less pressing.

Pest management in the past has largely concentrated on agricultural pests, and considerable expertise has been developed in that area. Thus, another question that must be examined is whether and to what extent the highly developed technology of agricultural pest management is applicable in urban settings. The issue of technology transfer must be considered in light of the heterogeneity of urban populations and the fact that urban pest management is far more likely to be carried out by individual householders than by trained applicators.

Given the variety of problems posed by pests in urban environments it is our task to examine what is known about the hazards posed by the pests themselves and by efforts to control them. The consequences of urban pest control, with special emphasis on the applicability of integrated pest management strategies to the urban environment, will be considered.

It is also important to examine the roles of federal, state, and local agencies in urban pest management and to determine how effectively they play them. This examination includes consideration of the effectiveness of laws and regulations and their current interpretation at all levels of government. An important issue that must be dealt with is whether there are legal constraints that inhibit effective urban pest management strategies. In many situations one must ask whether present legislation and regulation are sufficiently explicit to authorize effective pest management or, on the other hand, whether some of their specificity and detail interfere with the more flexible approaches that are urged from time to time. A recurring issue in this context is whether current law provides adequately not only for the management of pesticides but also for the management or control of pests, particularly urban pests. The role of different government agencies must be examined to determine whether it is necessary to assign general responsibility for urban pest management, particularly with respect to public health, to a specific federal agency. For instance, with the U.S. Environmental Protection Agency (EPA) carrying major responsibility for the management of pesticides, would it be reasonable to assign the obligation for the management of urban pests to a federal agency with major responsibility for the protection of public health, just as the management of agricultural pests has been assigned largely to the U.S. Department of Agriculture?

PERSPECTIVES ON URBAN PEST MANAGEMENT

The need to control pests in the urban environment has existed as long as villages and cities have. The increased density of human inhabitants, peculiar to the new pattern of life; the concentration of carbohydrate, lipid, and proteinaceous foodstuffs resulting from agricultural production and animal husbandry; the construction of dwellings for shelter; the accumulation of domestic animals and pets; and the problems of disposing of organic waste from high-density living patterns have provided a series of unique, urban ecological niches to be invaded and colonized by arthropods and vertebrates alike. With the development of human associations in urban areas, a distinct urban pest fauna has likewise developed. It includes the human body louse, the pubic louse, the bedbug, and distinctive urban species of fleas, flies, mosquitoes, triatomids, stored-product pests, and cockroaches. These have been joined along the evolutionary trail by the English sparrow, the domestic pigeon, and domestic rodents. The existence of this large group of human parasites and commensals provides ideal conditions for the transfer of pathogens to human beings.

Vector-borne human diseases have become significant during the last two thousand years and were responsible for the pandemics of the medieval era. However, better personal cleanliness, systematic disposal of organic wastes, and improved living facilities gradually stemmed the ravages of such diseases as plague and typhus until today these diseases are of minor importance. Other pest-borne diseases have succeeded them, however, each one characterized by a change in urban ecology. Filariasis caused by the nematodes *Wuchereria bancrofti* and *Brugia malayi* and transmitted to humans by the common mosquito *Culex pipiens* is the world's most prevalent urban disease, with an estimated 250 million cases yearly (Wright 1976). *Culex pipiens* breeds exclusively in highly polluted water and is invariably associated with inadequate disposal of human waste. Filariasis, a preponderantly urban disease, flourishes in the great metropolises of Southeast Asia where large human populations, lack of sanitary facilities, and heavy use of polluted water have caused the incidence of the disease to double in the past 20 years. The viral encephalitides transmitted between birds and humans by *Culex pipiens* also flourished as the urban environment became increasingly populated with domesticated birds. Increased transmission of hepatitis virus in the urban environment may well be due to vastly increased populations of the German cockroach, *Blattella germanica.*

Apart from their role as vectors of human disease, the presence of these stealthy domestic intruders causes annoyance and stress that exacerbate the tensions of modern urban living in the United States. Urban inhabitants have not yet resigned themselves to living in propinquity with lice, bedbugs, fleas, flies, mosquitoes, cockroaches, stored-product pests, rats, mice, sparrows, pigeons, and starlings. The widespread phobia about "creeping, crawling things" was intensified after World War II, when urban dwellers began to move to suburbia. There they encountered not only agricultural pests (flies from dairies and chicken ranches, and mosquitoes from irrigated farms), but also a host of new pests associated with trees, shrubs, lawns, ponds, and lakes: mosquitoes, yellow jackets, ants, subterranean and drywood termites, gypsy moths, Japanese beetles, mealybugs, scale insects, red spider mites, lawn moths, numerous other pests of plants, and a host of obnoxious and poisonous plants.

In 1946 it seemed as if simplistic measures might solve all these pest problems. In rapid succession modern technology produced DDT, lindane, and chlordane to kill lice, fleas, bedbugs, cockroaches, ants, dermestids, and termites; anticoagulant rodenticides to eliminate rats and mice; and 2,4-D to destroy unwanted weeds. The new pesticides supported a flourishing business of household pest exterminators and lawn care specialists who urged the application of more and more of the wonder chemicals. In addition, pesticides also became readily available at retail outlets; and information on their use was supplied by county, state, and federal extension agencies as well as by commercial advertising. Reliance on chemicals became so complete that lessons on the importance of good sanitation, so laboriously learned over the centuries, were neglected and ecological relationships were ignored. The pervasive philosophy became, "If a little pesticide is good, more will be better."

During the succeeding 30 years, however, it gradually became clear that "exterminating" certain urban pests with chemicals was less preferable than "managing" them in an ecological framework designed to minimize their numbers. One reason for this change in perception was the onset of pest resistance to pesticides, first experienced with DDT in 1946 and now widespread. Many insect and mite pests are now resistant to a variety of insecticides, as are rodents to the anticoagulant rodenticides and weeds to herbicides. The phenomenon of pest resistance gravely threatens the entire fabric of chemical pest control.

Another reason for reduced enthusiasm for chemical pest control was the development of a belated concern about the intrinsic wisdom of purposeful pollution of the environment through the application of pesticides. Homes are particularly vulnerable to unwanted side effects from chemicals used as fumigants, soil poisons, and residual sprays. These chemicals can enter the human diet directly, through food and water, or otherwise contaminate human surroundings. The suburbs, with their proliferation of private sanitary systems and canals that drain into larger streams and lakes, have proved highly vulnerable to pesticide contamination of water sources. The pollution of Lake Michigan with DDT from attempts to control Dutch elm disease and mosquitoes provides a particularly notable example. (The problems of pest resistance and urban pesticide use are considered in more detail in Chapters 2 and 3, respectively.)

One of the most apparently reasonable solutions to this dilemma—i.e., how to control pests by methods that will be continuously effective and economically and socially acceptable—is integrated pest management. IPM promises to substantially decrease pesticide use, preserve the effectiveness of essential pesticides, and decrease the overall social costs of pest control activities. (A discussion of the use of IPM in urban areas is presented in Chapter 3.)

DEFINITION OF URBAN

For the purposes of urban pest management the Committee defined "urban" to include both slums and ghettos and middle- and upper-middle-income areas, whether in inner-city or suburban locations. In accepting a broad view of "urban," the Committee's intent was not to blur distinctions or to avoid difficulties. But this broad definition recognizes that the character of urban areas differs significantly in different cities, and that within any one urban area there are significant contrasts in environmental and social conditions that affect the kind of pest problems and other environmental problems that must be faced. Recognition of a variety of pests and pest management problems should not result—and has not resulted—in a blurring of values or loss of priorities.

In inner-city slums, pests and pest management present crucial and urgent problems of public health. In suburban areas, pest problems are more likely to present aesthetic and economic issues. While pest-related public health problems of the inner city deserve priority, the protection of economic and aesthetic values is also a valid and significant consideration. Moreover, it has never been easy to draw clear boundaries between different parts of urban areas, which shade into one another with respect to population density, types of housing and other buildings, socioeconomic characteristics, and natural environment. Although certain problems, including pest problems, are more specifically focused in certain types of urban areas, they usually cannot be confined. Pest-borne illnesses travel from the slums to other parts of the city, and urban renewal efforts in one area are likely to disperse pest populations into other areas in search of food and harborage. Wood-destroying insects attack homes regardless of their location.

Given these considerations, it proved easiest to define "urban" by exclusion and by a number of pragmatic considerations. "Urban" clearly excludes all areas primarily devoted to agricultural pursuits—areas characterized by food or fiber monoculture or animal husbandry—and it also excludes areas characterized by open land and low population density. For working purposes, "urban" as defined in this report includes Standard Metropolitan Statistical Areas (SMSA), particularly core areas. In general, areas outside the core with a human population density sufficient to support a population of urban pests are also included. Lack of precision in defining urban areas, however, presented no problems because pest management data developed in the past have not distinguished "urban" from "nonurban" uses of pesticides, nor have they generally required application of a specific definition of "urban" or "urban area."

The Committee's approach was to collect data and other information from persons in metropolitan centers and to review the literature on pests commonly regarded as urban pests because of their habits, their habitat, or their interaction with human beings in nonagricultural settings. The emphasis is on public health pests, and the urban setting is the area where certain pests thrive because of the density of human population or because of other environmental configurations regarded as primarily urban. Many pest problems in the outer city and suburban setting are similar to those of the inner city, but attention is also paid to the management of destructive pests that thrive on lawns and ornamental plants and that are attracted to an environment characterized not by deprivation, but by affluence.

Finally, urban pest management must concern itself with certain types of commercial establishments commonly found throughout urban settings—restaurants; industrial establishments that process food, fiber, or cloth; and institutions like hospitals, nursing homes, and houses of detention—all of which attract pests for a number of reasons, including overcrowding and storage and preparation of large amounts of food.

CHARACTERIZATION OF URBAN PEST MANAGEMENT AND SOME OF ITS IMPLICATIONS

Historically, urban pests have generally been classified according to their significance either to public health or economics. More recently, a third class of pests with an impact on aesthetics has received increasing attention, and in this class it is also appropriate to consider pests whose biological activities or mere presence evoke negative emotional responses.

Pests that fit in the third category may be characterized as those that offend the aesthetic sense or emotional well-being of people. As might be expected, the degree to which people are affected will vary considerably, and this in turn is related to individual perception and past experiences. Many pests that generally cause a negative emotional response do not cause significant material damage, but attempts to manage or control them may cost money. It also appears that "aesthetically displeasing" pests may cause people to perceive themselves as being in a degraded environment, and this in turn may affect their ability to cope with life in general (National Institute of Mental Health 1972).

It is not always easy to classify a pest in one of the three categories named above. To many people, for example, mosquitoes and fleas are primarily a nuisance, but public health becomes a relevant concern when either insect carries pathogens. Some vertebrates such as squirrels or certain birds may be thought by some to be pests, but regarded by others as desirable under some circumstances. Weeds provide a classic example of difficulty in classification. Depending on viewpoint, a given weed can fall into any of the three categories. Furthermore, some urbanites may consider weeds a natural part of the urban environment (at least in certain types of gardens and parks) and therefore desirable.

These examples demonstrate that people interact with urban pests in a variety of ways. While some pests can be classified with little difficulty, others cannot be characterized according to a simple scheme. The extent of many pest problems clearly depends on individual perceptions. It is logical, therefore, that cultural patterns and social, economic, political, and other factors be investigated before certain pest problems can be intelligently analyzed and managed.

Perhaps because the field of pest management has had its beginning in the agricultural setting, many professional pest managers have considered most urban pest problems as basically economic in nature. The choice of control methods has been influenced more by immediate economic goals and narrow objectives than by broader environmental, social, or public health concerns. Although environmentalists and some public health workers have urged a broader approach, the traditional view is still widely held among researchers and extension service personnel. It is becoming increasingly clear, however, that this viewpoint is too narrow in light of the great diversity and complexity of pest problems in urban areas.

In the opinion of the Committee, pest management techniques drawn from public health, agricultural, and other pest control efforts including physical, chemical, and environmental manipulation and other methods should all be used under various conditions and combined in whatever manner promises to have a maximum effect upon the target organism and minimum adverse effect upon human beings and the natural environment, including nontarget organisms.

One practical way to deal with pest problems is to cast them in an ecological framework in which the relationship of the pest to people, to its physical environment, to its food sources, to other animals and plants, and to man-made structures can be assessed. This approach, which is similar to that proposed by Olkowski et al. (1976) and Frankie and Ehler (1978), allows one to assess the symptoms of the problem as well as its basis. It is only logical to assume that a more complete understanding of a pest problem will lead to a more rational, realistic, and effective solution.

Pest management and control in urban areas is practiced by a wide variety of persons. At one extreme are sophisticated management programs operated by teams of highly trained professionals. At the other are simple management efforts undertaken by untrained and overzealous urbanites. In between lie combinations of professionals, paraprofessionals, and urban residents.

Regardless of the level at which pest management is practiced, however, there is a continuing need for education of both pest managers and recipients of their services. Until recently, education has been given less than adequate attention, mainly because of cost and difficulties in information transfer. In addition, adequate biological, behavioral, and ecological information about many urban pests is lacking. For pest management to be effective there must be a concerted effort to educate all affected parties. Education of urbanites, for example, leads not only to an understanding of the pest and the general problem it causes but also to greater involvement on the part of residents in solving the problem. This is extremely important when cultural or habitat modifications are required as tactics in a management scheme.

CONTRASTS BETWEEN AGRICULTURAL AND URBAN PEST MANAGEMENT

There are a number of basic similarities between agricultural and urban pest management. Ecological relationships are important in both, and quite often the problems in both areas have a clear economic base. Many of the specific tools used to suppress pest populations in agricultural and urban environments are the same. In the case of pesticides, for example, most of the chemicals used in urban environments were first developed for use in the agricultural sector, where the market is large enough to justify development. Finally, the philosophy of integrated pest management is applicable to both areas.

Careful examination of urban pest management, however, also reveals differences between it and agricultural pest management. Unlike in urban areas where pest management is often a mere incident of daily life, pest management in agriculture is an integral part of the task of agricultural production. In agriculture and particularly in monoculture, it is meaningful to speak of managing pest populations until they are at levels below those that cause an economic loss larger than the control costs, i.e., levels at which incremental savings would be less than the additional cost required to achieve them. In urban areas the emphasis may be on management or on "eradication," depending upon whether the fundamental problem is a matter of economics, aesthetics, or public health. It is important to note that all the problems caused by urban pests have economic aspects, but only some lend themselves to economic decision making.

Furthermore, the training and background of agricultural pest managers may be different from that of urban pest managers. The agricultural pest manager is often a trained professional, whereas in the urban sector the pest manager may be a professional, a paraprofessional, or an urban dweller with no formal or informal training. From both the theoretical and practical standpoints, it is easier to transfer new technology to the agricultural pest manager than to a member of a heterogeneous urban population who is not using the information to make a living. Thus, the educational challenges in urban pest management are greater.

Finally, intelligent management of a particular pest in an urban environment frequently involves consideration of other than biological factors. The pest and its management must often be viewed within the context of a particular socioeconomic and political setting. Understanding human attitudes and behavior becomes of the utmost importance, since many urban pest problems involve problems of individual and social psychology. Such considerations generally do not apply in agricultural pest management.

EDUCATION AND URBAN PEST MANAGEMENT

Formal and informal education in urban pest management varies, depending on the needs and interests of those involved and the requirements of educational institutions and regulatory agencies. University and extension service clinics are held throughout the country to inform the general public about pests and ways to manage or control them. A variety of formal and informal training courses is available throughout the United States for professional and paraprofessional pest managers, sanitarians, and pest control operators.

At the university level only a minor effort has been made, to date, to educate undergraduates in the broad aspects of urban pest management. It appears that most universities with urban-oriented curricula do not require course work that exposes students in urban entomology and pest management to the physical, social, economic, and political environments of humans in which pest organisms are often involved. Instead, such students are frequently placed in agricultural pest management courses with the expectation that they will be able to apply their knowledge to urban pest management.

Some innovative educational efforts, however, did come to the attention of the Committee. For example, several managers of mosquito abatement districts in northern California have made recommendations to city planners and real estate developers regarding mosquito problems that may accompany proposed land developments. Through simple modifications of construction plans, these managers have been able to demonstrate how to anticipate and reduce mosquito problems. This effort has resulted in significant savings by eliminating or reducing the need for mosquito control following construction. Unlike many pest management efforts, it is aimed at controlling the basic source of the mosquito problem, not its symptoms.

CRITICAL TOPICS IN URBAN PEST MANAGEMENT

Effective, responsible management of urban pests must take into account a variety of issues, some of which have not previously been raised or examined in the context of urban pest management. No predetermined group of concerns will fit every circumstance, but the areas addressed below should be considered when a comprehensive assessment of the impact of urban pests and of strategies for managing them is undertaken.

HAZARDS OF PESTS IN THE URBAN ENVIRONMENT

The most significant adverse effect of pests is pestilence—human disease. There is a chance for the occurrence of disease wherever people are exposed to insects or animals that bite, suck blood, sting or otherwise puncture the skin, or deposit their feces, urine, saliva, or other secretions on human food or water or in places where they can be inhaled. When human beings live in proximity to animals that harbor infectious agents that also can infect humans, there is danger that the agents will be transmitted. The presence of potential arthropod vectors enhances this risk.

Recently, pest-borne diseases have caused epidemics of staggering proportions in Europe, Asia, and Africa. The precise conditions for such pandemics do not now exist in the United States. Nevertheless, the danger that significant outbreaks of pest-borne diseases may occur in American cities and towns is never entirely absent.

Outbreaks of up to several hundred cases of mosquito-borne encephalitis, with average mortality rates of more than 5 percent, occur almost annually in various parts of the nation (U.S. PHS 1968-1977). Mosquito vectors of dengue, yellow fever, and malaria—all diseases to which human beings are highly susceptible—are widespread in this country. Each year hundreds of travelers or immigrants with infectious malaria or dengue, and a few from areas where yellow fever is prevalent, enter this country. Under appropriate circumstances they could introduce those diseases here.

Plague is widespread among the wild rodents of the West that live in close proximity to domestic rodents and fleas in urban and suburban dwellings. Prior to 1910 there were more than 100 reported cases of human plague each year, but since 1970 the number of cases has been about 15 to 20 (Anderson 1978). Nevertheless, the potential for a major outbreak still exists. In all parts of the country, especially the East and Southeast, more and more suburban dwellings are located on the fringes of woodlands, close to populations of raccoons, rabbits, woodchucks, skunks, opossums, field mice, and other vertebrates, many of which feed on human garbage and other human food sources. These animals are infected with various microorganisms—viruses, rickettsia, bacteria, and protozoans—and some of their diseases (rabies, tularemia, Rocky Mountain spotted fever) are transmitted to people by ticks, mosquitoes, or the animals themselves. New or previously unrecognized tick-borne diseases with animal reservoirs (Lyme arthritis, babesiosis) recently have been identified in people living in suburban areas (Anderson et al. 1974, Healy et al. 1976). There is a possibility that one or more of these hitherto wild microorganisms might become adapted to new vectors and to human beings—just as typhus and plague apparently did several centuries ago.

Many people are widely exposed to arthropods and vertebrates that are potential carriers of epidemic disease or causes of infestation or annoyance. An estimated 3 million Americans (chiefly children and adolescents) are infested with head lice, and probably an equal number have pubic lice (Elzweig and Frishman 1977). Millions of dwellings contain cat and dog fleas, and the number of dwellings with mice is probably equally great. Norway rats are still present in a large number of cities, and black rats, or roof rats, are important in the South and West. A wide variety of mosquitoes is present in and around urban areas, as are the pigeons and starlings that serve as reservoirs of some mosquito-borne diseases.

Pest-borne diseases and attacks by pests themselves account for 100 to 300 deaths in the United States annually, for probably 20 times this number of serious disabling illnesses and injuries, and for several million episodes of illness or injury that are serious enough or painful enough to require medical attention. Annual surveillance reports from 1968 to 1977 list about 380 cases of laboratory proven mosquito-borne encephalitis causing 20 deaths (U.S. PHS 1968-1977); about 1000 cases of Rocky Mountain spotted fever causing over 40 deaths (U.S. PHS 1979b); and sporadic cases of plague, murine typhus, rabies, and rat-bite fever (Pratt et al. 1976, U.S. PHS 1979a). Each year more than a million animal bites require medical treatment. Approximately 800,000 of these are attributed to dogs, some of which are strays but most of which are free-ranging pets. There are more than 95,000 cat bites and 43,000 rat bites, as well as 30,000 bites from other animals including skunks, bats, and raccoons (Moore et al. 1977). At least 11 fatal dog-bite attacks were documented in 1974 to 1975 (Winkler 1977). Many of the rat bites cause disfiguring injuries of infants, and some are fatal. Approximately 30,000 people have to undergo prolonged, painful, and sometimes dangerous treatment for rabies each year because of animal bites (Winkler and Kappus 1979).

The feces and urine of dogs, cats, rats, and mice, if deposited in places where they can come into contact with people or contaminate human food and water, can transmit various kinds of infectious diseases such as leptospirosis, and infestations, e.g., of various tapeworm larvae including one (visceral larval migrans) that damages the retina of the eye and is probably the leading cause of severe retinal disease in children (Pollard et al. 1979). A variety of fungal organisms grow in environments contaminated by pigeon and starling feces and the air-borne spores cause systemic infection in several hundred people each year, with more than 100 deaths (Fraser et al. 1979). Human reactions to envenomization (the stings and bites of wasps, which include hornets and yellowjackets, and of bees, spiders, scorpions, and snakes) account for approximately 40 deaths and perhaps more than 20 times this number of serious disabling injuries each year (Barnard 1973). Infestations of people with itch mites, lice, and fleas; of homes and other buildings with bedbugs, fleas, flies, and cockroaches; of both urban and suburban areas with mosquitoes; and of suburban areas with ticks, fire ants, chiggers, and similar organisms cause pain, discomfort, and the threat of illness to millions of people throughout the country every year.

Many of the estimated figures given above are probably low, and it is likely that they would be doubled if there were more accurate and complete reporting (see Chapter 2).

HEALTH CONCERNS ARISING FROM URBAN PESTICIDE USE

Although the initial use of chemicals to manage or control pests led to effective control of pest-borne diseases and the elimination of certain zoonotic reservoirs, subsequent detection of adverse health effects from pesticides soon tempered the initial enthusiasm for sole reliance on this pest management strategy. Widespread human poisoning from pesticides, the effects of chronic and incidental exposure, and increasing pest resistance to chemicals are currently the major areas of concern.

Since the 1960s, when organochlorine pesticides were replaced by faster-acting but more toxic organophosphate and carbamate insecticides, overexposure has resulted in an undetermined number of cases of systemic poisonings (Davies 1977). The effect of these pesticides on human beings can be seen most clearly in those parts of the world where knowledge of safety precautions has not kept pace with chemical technology (Davies et al. 1978), but even in the United States—where much is known about pesticide safety, overuse, and misuse—there is a large number of poisonings (see Chapter 3). Both systemic and topical effects can be seen (Maddy 1976), but these effects are poorly documented in states other than California, and the true magnitude of the problem is unknown.

Medical awareness of the effects is becoming increasingly sophisticated, and there is recognition that such conditions as male sterility (Whorton et al. 1977), neurologic and renal diseases (Sanborn et al. 1979, Mann et al. 1967), cancer (A. Baetjer, M. Levin, and A. Lilienfeld, Allied Chemical Company, personal communication, 1978), teratogenic effects (Field and Kerr 1979), and behavioral disorders (U.S. Congress, Senate 1976) are sometimes related to pesticide exposure. The medical significance of chronic and incidental exposure has become as great as that of acute exposure.

Another problem related to public health is increased vector resistance to pesticides, such as has been encountered with carbamate and organophosphate insecticides in the anopheline mosquito (Georghiou 1972) and with warfarin in the rat (Jackson et al. 1971, 1975). The resistance of various pest species to chemical agents has led to a need for strategies that do not use chemicals as the sole mechanism of control.

Concern about the effects of pesticide exposure on human health was first expressed in agriculture. But it is now presumed that significant exposure is also occurring in the urban setting, where the population at risk is not only different, but also larger. In addition, the chemicals used by urban residents are often different from those used in agriculture, and misuse is frequent. Unfortunately, however, health concerns about the effects of pesticide use in urban areas are based largely on effects documented in agricultural situations, and few if any comprehensive studies of the effects in urban areas have been initiated (see Chapter 3).

URBAN PESTS AND ENVIRONMENTAL POLICY

The presence of pests in urban environments is a significant factor in the decline of the nation's cities and the overall quality of life experienced by a large portion of urban dwellers. To date, however, the management of urban pests has received remarkably little attention in environmental analysis and policy, and individual perceptions of pests and their impact have only recently received attention (e.g., Frankie and Levenson 1978, Levenson and Frankie, in press). Inventories of urban pests and research on their behavior and their impacts on human health and urban economics are scanty. A recent study by Nelson (1978) attempted to put in perspective some of the problems of pest impacts on urban quality of life. Nelson noted that Surtees (1971) defined urbanization as the replacement of a natural ecosystem by a dense center of activity created by humans, containing humans as the dominant species, and environmentally organized for the survival of humans. Urbanization may thus be viewed in an evolutionary sense as one of the selective forces that determine the extent of occurrence and persistence of medically important arthropods.

Despite the lack of scientific research on urban pest problems, however, there is increasing evidence that pests exert a serious and degrading impact on community well-being in urban areas.

Meanwhile, environmental analysis and policy have developed without the benefit of a basic framework for considering the unique characteristics of the urban setting. Historically, environmental research has focused on air and water quality and land management. Concern with individual environmental abuses has overshadowed the importance of understanding the interaction of such abuses in a complex environmental system—natural and man-made—and blurred the economic and social impacts. Environmental research has ignored the critical need for basic inventories and baseline data on a variety of other environmental problems in urban areas. Furthermore, urban policies on such matters as housing often fail to include environmental issues in their consideration of neighborhood maintenance problems. The low status assigned to urban pest management research reflects the current state of both environmental research and urban-related policies. Not surprisingly, research and demonstration projects that might lead to improved strategies for the use of such approaches as integrated pest management are missing from most municipal pest management guidelines.

Analysis of Urban Environmental Problems

As mentioned above, current environmental research focuses on defining the special problems associated with specific pollutants, particularly air and water pollution, noise, solid waste, and pesticides. Assessment of the impact of these pollutants is stressed, mainly in terms of the health and economic costs to individuals and society, either in the workplace or in the open environment. There is an emphasis on objective, quantified indicators (i.e., scientific measurements) and some analysis of perceptions. Although there has been some success in isolating particular pollutants and in assessing their impact on morbidity and mortality, these analyses suffer from the bias and fragmentation of specialized disciplines and from the broad, aggregative approaches used in impact assessment. Aggregate analyses, in particular, fail to isolate the negative contributions of such factors as housing conditions and densities, nonresidential activity mixes, and other physical characteristics of urban environments. The aggregative approach also assumes that the impacts of pollution are similar throughout the urban setting, irrespective of income, social position, or geographical location. Core-city residential neighborhoods, central business areas, industrial suburbs, and fringe residential suburbs are often merged in such analyses, as are the population groups in these settings. Thus, governmental policies and actions based on aggregate impact analyses often understate the social costs of pollution and the benefits that accrue from improvements (U.S. EPA 1971).

Limited evidence, provided by epidemiological studies and inferences about the distribution of air pollution, suggests that the intensity of environmental problems differs considerably among different types of urban settings. The most severe problems are typically found in older, high-density settings like the central-core cities of major metropolitan areas (Berry and Horton 1974, Goldman 1970, McCaull 1976). Moreover, building density, frequently highly correlated with occupancy density and poverty in cities, further compounds the problems of exposure and health risks. By contrast, sometimes serious but localized problems tend to occur in less dense settings and even in newer suburban communities (e.g., exposure to herbicides).

Another problem involves environmental maintenance activities. Unlike the natural environment, which often is able to adjust itself to stress, urban settings require deliberate and sustained maintenance of buildings and infrastructures. Where age and reductions in maintenance are found together, neighborhood effects may be severe.

Understanding urban environmental problems also involves understanding the interactions between stressed environments and human social and economic activities. The health and economic well-being of disadvantaged people—particularly the very old, the very young, racial minorities, and the handicapped—may be more seriously affected by environmental stress than are the health and economic well-being of the less disadvantaged.

Although the higher concentration of human activities, the aged infrastructure and buildings, and disadvantaged populations generally combine to create visible environmental problems, such problems are not limited to older core cities. They are found throughout more broadly urbanized and suburban settings. This is true of pest and pest control problems as well as environmental problems in general. Thus, ground- and surface-water pollution or solid-waste management practices can create such major problems as vermin and mosquito infestation for whole regions (Nelson 1978).

A selected bibliography on urban environmental research and policy is presented in Appendix C.

Pest Problems and Perceptions of Them

"Quality of life" is a term increasingly used in attempts to measure nonquantifiable values—that is, the complex set of values associated with human needs, demands, and expectations. Efforts have been made to relate the quality of life in urban environments to some objective measure of well-being using quantifiable indicators related to amenities, open space, and cultural opportunities, or to such less quantifiable, or descriptive, indicators as "satisfaction" with economic and social status and neighborhood.

In current research, definitions of quality of life are approached through two types of inquiry. One centers on "objective" measures of environmental quality against which cross-sectional comparative data can be used to construct indices or scales of relative attainment. The other approach attempts to measure the subjective preferences, perceptions, and satisfaction of individuals with environmental values.

In the first type of inquiry, statistics on physical aspects of the environment—open space, housing characteristics, and housing values, along with income, social and occupational status, state of health, and disease rates—are used to construct indices of the relative quality of life attained by local communities or neighborhoods (Allardt 1975, Hall 1976, Andrews and Withey 1976, Campbell et al. 1976, Hankiss 1978). Flax (1972) examined the relevance of selected indicators from a number of major metropolitan areas as measures of qualitative values. Data on employment rates, household income, costs of housing and transportation, death rates (including suicide), educational levels, drug addiction, and charitable contributions were analyzed as proxies for subjective categories, such as poverty, health, public order, racial equality, air quality, citizen participation, and social disintegration. While generalizations about the status or ranking of communities can be made this way, the weighting systems used and the meaningfulness of the variables as measures of qualitative values have aroused considerable criticism.

The many objective values that can be measured—such as cultural resources, parks, open land, or even air pollution or pest concentrations—are not strictly comparable in nature. Thus, deciding whether large numbers of a particular pest are "worse" or pose a greater threat to health than a high concentration of sulfur dioxide is not possible, even if quantifiable.

Dissatisfaction with objective indicators has led to increased interest in subjective preferences and perceptions. Milbraith (1979) examined some of these objections and concluded that while objective indicators can be used to draw reasonably valid inferences about environmental conditions, definitions of quality of life should take into account perceived values as well. Several studies have shown a low correlation between objective measures of a condition and subjective measures of the way the condition is perceived by different population subgroups. Quality is necessarily a subjective concept, and inferences may be drawn most validly by asking individuals to report their subjective feelings (Milbraith 1979). This may be particularly important for poor people and the disadvantaged whose views are rarely sought and often misunderstood.

Problems may also be encountered in arriving at accurate statements of subjective preferences. As Milbraith (1979) reveals, the most serious problem is the possible theoretical bias that, for policy purposes, "preferences" are equivalent to "social goals." The danger of Milbraith's subjective approach is that lowered expectations could support a dangerously low health standard in an urban slum, for example, merely because the inhabitants do not "express" a preference for purer water or cleaner air. A second problem that may arise in the measurement of attitudes is that investigators may be unable to pose questions in ways that elicit accurate statements of perceptions.

Urban Pest Problems and the Man-made Environment

Whether measured in objective terms or in terms of perceptions and preferences, there is a continuum of environmental abuse in urban areas ranging from most severe (core-city slums) to least severe (affluent fringe areas), with a full range of conditions possible between (Nelson 1978). It is therefore useful to examine the causes of present urban environmental conditions before assessing the particular case of urban pests.

The physical decline of previously sound housing has taken place at an alarmingly rapid rate in the United States, particularly in older, inner cities. In many major cities the rate of property disinvestment, abandonment, and lowered maintenance standards has escalated in recent years. The 1974 "Annual Housing Survey" of the U.S. Bureau of the Census and HUD (U.S. HUD 1977) indicates the extent of the problem. The survey found that nearly 10 percent of the occupied housing in the United States had two or more serious structural defects (see Table 1.1) and that the health impact of as many as 8 percent of those with three or more defects could be considered severe by national standards. The Harvard and MIT Joint Center for Urban Studies estimated that over 20 percent of the nation's 5 million defective housing units would be concentrated in core cities and that 50 percent of the low-income households in these areas would be living in deteriorating housing by 1980 (Frieden and Solomon 1973). The studies indicate that the problem is compounded by (a) declining growth in income relative to housing costs, resulting in increases in the proportion of the population paying in excess of one fourth of family income on housing; and (b) growth in occupancy rates and crowding in standard housing in cities (Frieden and Solomon 1973).

Housing deterioration and slum creation are important factors in the decline of urban environments. Dilapidated buildings are prime sites for

TABLE 1.1 Occupied Housing Units with Specified Defects and Incidence of Multiple Defects, 1974

Number/or Type of Defects	Total Units in Group (in thousands)	Percent of Total Stock	Average Income of Households ($)
None	40,827	57.6	13,300
One	18,668	26.4	12,600
Electrical	1,658	2.3	10,500
Kitchen	218	0.3	7,700
Plumbing	3,303	4.7	12,000
Bathroom	166	0.2	8,800
Heating	1,810	2.6	14,300
Roof or basement leaks	6,030	8.5	14,300
Walls, floor or ceiling cracks or holes	2,960	4.2	10,700
Vermin	2,523	3.6	12,400
Two	7,193	10.2	10,700
Three	2,564	3.6	9,500
Four	1,047	1.5	7,300
Five	407	0.5	6,300
Six	119	0.2	4,700
Seven	5	0.0	5,400
Eight	0	0.0	–

SOURCE: U.S. HUD (1977).

pest infestation and fire. Deterioration of public services, particularly sanitation, exacerbates the problems, and the presence of the poor reinforces them. According to Nelson (1978):

These man-made conditions provide food, water and harborage for the arthropods or the hosts of these arthropods. Although urbanization may produce an increase in certain natural conditions that are necessary for survival of the arthropod or host components of the biocenose, it more often produces artificial and simulated conditions that allow for development of unusually dense populations. *Urbanization selects those arthropods that will succeed in urban environments by offering substitutes for their natural requirements.* (Emphasis added.)

Current data on housing and environmental conditions suggests that certain pest species, particularly vermin, constitute a major and growing urban problem. According to the 1974 "Annual Housing Survey" (U.S. HUD 1977), approximately 3.6 percent of all housing units in the United States and nearly 13 percent of the defective units showed signs of vermin. Vermin ranked second among the 18 specific defects cited in the survey.

The most serious problem, however, may be the upward trend. While all other defects declined between 1973 and 1975, the number of units affected by vermin problems increased 17 percent, from 6,676,000 to 7,836,000.

Current evidence also suggests that a significant part of the problem is concentrated in inner cities. Unpublished data on rat bites and vermin complaints from New York City, reviewed as part of this Committee's work, indicate a high correlation between the presence of vermin, older housing, and poverty in declining high-density areas of Harlem, the Lower East Side, central Brooklyn, and the south Bronx (Rafi Al-Hafidh, New York City Department of Health, personal communication, 1979). The Council on Environmental Quality (1971) estimated that 60 to 80 percent of the rat bites reported annually in the United States occurred in inner-city neighborhoods.

Growth in the number of urban pests can be attributed to several factors, including lack of building maintenance, declining public services (e.g., waste disposal) and enforcement of health and building codes, and poor household habits. Households, particularly those of the poor and newcomers from rural areas, may not understand the potential dangers of urban pests. Lack of education or sophistication, or fear, may result in reluctance to report negligent landlords or the absence of garbage collection to municipal authorities. Building and site owners, while legally responsible for maintaining structural and environmental standards, frequently ignore their responsibilities. Crevices and holes in buildings, inadequate receptacles for household waste, and sloppy housekeeping in hallways and basements, on roofs and in yards may result in accumulated debris and harborage for pests. The large and increasing number of abandoned buildings and vacant lots also provides opportunities for pest propagation. Finally, public pest management methods may have contributed to the problem. An example is the sole reliance on certain toxic chemicals without other techniques like habitat modification; little interest has been shown thus far in integrated approaches that take advantage of natural, biological, and other environmental controls like the breeding of pest-resistant plants or introduction of pest predators.

Environmental degradation in the cities has been abetted by reduced capital investment, lowered maintenance, and inadequate enforcement of standards. The reduction in investment in sanitary (sewer and water) systems, transportation, and public buildings—including group housing facilities like prisons and hospitals—has been particularly significant (Bahl et al. 1978). Expenditures for weekly household waste removal, public site maintenance, and street litter removal have declined relative to the growth in waste volume. Household waste generated during the past decade is estimated by EPA to have grown by 700 percent, for example, while municipal expenditures for waste collection and disposal have increased by less than 20 percent (CEQ 1977).

How urban dwellers perceive pest problems is crucial to public decisions about the degree of financial commitment to these problems, but there is no agreement about public perceptions. Some investigators have found that strength of feeling about environmental hazards is correlated with income, race, and social status. A study in Los Angeles, for example, showed that the poor and racial minorities perceived less environmental danger than the more affluent (Van Arsdol et al. 1964). The difference, it was suggested, resulted from the tendency of the poor and minorities to place higher priority on needs other than environmental quality. Other studies, however, present less conclusive evidence. Housing satisfaction surveys of urban dwellers undertaken during the 1960s disclosed major complaints about environmental conditions, including lack of sanitation, presence of vermin and pests, abandoned buildings, inadequate municipal sanitation services, and excessive noise and other traffic impacts (Walter et al. 1979.) A recent study in the District of Columbia (Woody et al. 1980) found that neighborhood leaders ranked environmental problems fourth among their neighborhood's 10 most important problems, while rankings drawn at citizens' meetings placed "environmental abuses and services" third, following "housing" and "community participation in government decisions."

Suburban residents, moreover, do not always show greater perception of environmental hazards. Studies in Los Angeles, for example, suggest that suburbanites fail to perceive accurately threats such as floods and violent storms (Van Arsdol et al. 1964). The hazards posed by the use of pesticides in homes and on lawns and gardens and by low-grade sewage disposal arrangements are also often poorly perceived by the more as well as the less affluent.

SOCIAL AND PSYCHOLOGICAL CONCERNS OF URBAN PEST MANAGEMENT

Social and psychological factors are critical in developing effective urban pest management programs. Sanctions and controls on pesticide use by manufacturers, governmental agencies, retail merchants, family, and friends all lead to social actions with their own norms, expectancies, and guidelines. In addition, "pest" is a subjective term. An organism is not a "pest" by itself but only becomes one when people define it as one because of their attitudes, perceptions, or values. Thus, study of the social-psychological determinants of pest management practices and their consequences is crucial to an understanding of current and future pest control techniques and programs. Despite scientific interest in generalized attitudes and behavior (some of which was noted in the previous section), data on urban dwellers' perceptions of pests and pesticide use and on how they are affected by pests and urban pest management practices are for the most part unavailable.

Social-psychological determinants of urban pest management behavior include such factors as culture, attitudes, personality, values, and norms. As noted earlier, the consequences of pests and of pest management techniques in urban areas are different from those in agricultural settings. In agricultural pest management, economic decisions are relevant and usually predominant. In the urban environment, however, abstract concepts like "aesthetic need," "nuisance value," and "quality of life" seem more salient. Social impact analysis has emerged as a way of conceptualizing and measuring the social consequences of various environmental changes.

Consideration of social-psychological factors is also critical in developing integrated pest management (IPM) techniques and evaluating their effectiveness. For example, while it is not often explicitly delineated in writings on IPM, altering people's conception of "pests" may allow for their effective management. In effect, by changing what is perceived as a pest, or an aesthetic injury, or a nuisance, one controls the pest by controlling the perceiver.

INTRODUCTION TO PEST SURVEYS

Urban pests include a wide variety of vertebrates (rodents, birds, bats), invertebrates (insects and other arthropods), weeds, and wood-destroying fungi (see Chapter 2 and Appendix B). Except as they are carried by these pests, microorganisms pathogenic to human beings and animals were excluded from consideration in this study. Plant nematodes were not considered, but other plant pathogens, because of their relevance to lawn and garden environments, were included (see Appendix B).

SCOPE OF STUDY

METHODS OF MANAGING URBAN PESTS

Various methods have been used to manage or control urban pests. Because of their long history of use and misuse, pesticides are given considerable attention in this report (see Chapter 3). The most commonly used chemicals are grouped according to general class, and their target pests, target sites, and frequency of use are described. The amounts of pesticides entering the urban environment also are examined. Limited information was gathered on the geographical distribution of pesticide use in the United States and, where available, in selected states or regions. An attempt also was made to explore use patterns within localized areas by particular socioeconomic groups, neighborhoods, and individual residents.

Pesticides were examined from the standpoint of their effectiveness on target species, both as used in programs that are primarily pesticide-based as well as in more integrated programs. Attention was also paid to the ever-increasing phenomenon of pest resistance. Pesticides were also examined from the standpoint of their effect on nontarget organisms, particularly their effect on human beings. Information from clinical and epidemiological sources was sought in order to assess the impact of formerly and currently used chemicals on human health (see Chapter 3) and on vector-borne diseases of public health concern (see Chapter 2). Selected data on human exposure to pesticides used in urban pest management programs were reviewed, and both the direct effects of acute and chronic exposure and the indirect effects of pest resistance on health are discussed. The adverse effects of pesticides on wildlife and plant life in urban areas were also examined. A discussion of the socioeconomic and political implications of pesticide use and misuse is presented (see Chapter 4). Finally, present and possible future trends in pesticide use in urban areas were explored.

Use of controls other than (or in addition to) standard pesticides were examined by the Committee (see Chapter 3). These included relatively simple controls, such as habitat modification (including sanitation), cultural controls, biological controls, the use of resistant host plants, and more sophisticated methods like insect growth regulators and genetic manipulation.

Use of an integrated approach in urban pest management, as exemplified in residential and commercial facilities, was also examined (see Chapter 3). As with the standard pesticides, these methods were evaluated from the standpoint of their effect on target and nontarget organisms in the urban environment. In addition, these methods were examined in light of relevant socioeconomic and political considerations. Where appropriate, further investigation of methods that show potential usefulness in future pest management schemes is suggested.

Experimental and Operational Urban Pest Management Programs

The Committee evaluated selected experimental and operational pest management programs in the United States (see Chapter 3). This was not an easy task, since these programs range from the most sophisticated solo and team efforts by professionals to simple pest management programs that are being carried out by paraprofessionals or untrained individuals. The following sources were surveyed for information:

1. The existing literature;
2. The extension services of land grant institutions with an interest in urban pest management;
3. Researchers in the field;
4. City, county, and statewide programs in selected states;
5. Selected public health agencies;
6. The Center for Disease Control of the U.S. Public Health Service; and
7. Selected commercial operators and corporate programs.

The tactics of each urban pest management program were examined. When available, data on the efficacy of the program on target pest populations were gathered. The Committee also attempted to gather information on the impact of various programs on human health and nontarget organisms. If there were other environmental effects associated with these programs, they also were assessed.

NOTES

1. The first federal pesticide legislation was the Insecticide Act of 1910, 36 Stat. 335. For historical developments up to the passage of the present Federal Insecticide, Fungicide, and Rodenticide Act as Amended, Public Law 92-516, 86 Stat. 973, and beyond, see F. Grad (1979) *Treatise on Environmental Law*, Sec. 8.02 [1], New York: Matthew Bender Company.

2. For an account of air pollution legislation with special urban emphasis, see Grad, note 1 above, Sec. 2.03 [1] and Sec. 2.04 [2]. For impact of lead emissions on inner-city areas, see U.S. Congress, Senate (1972) Hearings, Subcommittee on the Environment, Committee on Commerce, 92nd Congress, 2nd session. See also E. Hall (1979) Pages 24-26, *Inner City Health in America*, Washington, D.C.: Urban Environment Foundation; National Research Council (1979) *Lead in the Human Environment*, Washington, D.C.: National Academy of Sciences.

3. *See* Federal Water Pollution Control Act Amendments, Sec. 307, 33 U.S.C. Sec. 1317, and Safe Drinking Water Act, Public Law 93-523, 88 Stat. 1660, 42 U.S.C. Sec. 300h-l.

4. *See* Olkowski et al. (1978) *Urban Pest Control in California: An Assessment and Action Plan*, Sacramento: California Department of Food and Agriculture, Contract No. 9958; and Hearings, note 2 above, pages 4-5.

5. The Congressional Declaration of National Housing Policy in the 1949 Federal Housing Act stated that "the goal of a decent home and a suitable living environment for every American family . . . " 42 U.S.C. Sec. 1441. See also National Commission on Urban Problems ("Douglas Commission"), *Building the American City*, 91st Congress, 1st session, House Document 91-34.

6. For example, U.S. EPA (1971) *Our Urban Environment and Our Most Endangered People*, Task Force on Environmental Problems of the Inner City, pages 5-6. Washington, D.C.: U.S. Environmental Protection Agency; *see* also Hall, note 2 above, pages 27-29.

7. *See* for example, Hall, note 2 above, pages 27-29; Senate hearings, note 2 above, pages 2-13 and 236-237. *See* also U.S. EPA, note 6 above, pages 5-6 and 8-14.

REFERENCES

Allardt, E. (1975) Dimensions of Welfare in a Comparative Scandinavian Study. University of Helsinki Research Group for Comparative Sociology, Research Report No. 9. Helsinki, Finland.

Anderson, A.E., P.B. Cassady, and G.R. Healy (1974) Babesiosis in man. Sixth documented case. American Journal of Clinical Pathology 62(5):612-618.

Anderson, E.T. (1978) Plague in the continental United States, 1900-1976. Public Health Reports 93(3):297-301.

Andrews, F. and S. Withey (1976) Social Indicators of Well-Being. New York: Plenum Press.

Bahl, R., B. Jump, Jr., and L. Schroeder (1978) The outlook for city fiscal performance in declining regions. *In* The Fiscal Outlook for Cities, edited by R. Bahl. Syracuse, N.Y.: Syracuse Univerity Press.

Barnard, J.H. (1973) Studies of 400 Hymenptera sting deaths in the United States. Journal of Allergy and Clinical Immunology 52:259-264.

Berry, B.J. and F.E. Horton (1974) Urban Environmental Management. Englewood Cliffs, N.J.: Prentice Hall, Inc.

Campbell, A., P. Converse, and W. Rogers (1976) The Quality of American Life. New York: Russell Sage Foundation.

Council on Environmental Quality (1971) Environmental Quality. Second Annual Report. Washington, D.C.: U.S. Government Printing Office.

Council on Environmental Quality (1977) Environmental Quality. Eighth Annual Report. Washington, D.C.: U.S. Government Printing Office.

Davies, J.E. (1977) Pesticide management safety—from a medical point of view. Pages 157-167, Pesticide Management and Insecticide Resistance, edited by D.L. Watson and A.W.A. Brown. New York: Academic Press.

Davies, J.E., R.F. Smith, and V. Freed (1978) Agromedical approach to pesticide management. Annual Review of Entomology 23:353-366.

Elzweig, M.A. and A.M. Frishman (1977) The role of the PCO in controlling human lice. Pest Control 45(11):33-34.

Field, B. and C. Kerr (1979) Herbicide use and incidence of neural-tube defects. Lancet 1(8130):1341-1342.

Flax, M.J. (1972) A Study in Comparative Urban Indicators: Conditions in 18 Large Metropolitan Areas. Washington, D.C.: The Urban Institute.

Frankie, G.W. and L.E. Ehler (1978) Ecology of insects in urban environments. Annual Review of Entomology 23:367-387.

Frankie, G.W. and H. Levenson (1978) Insect problems and insecticide use: Public opinion, information and behavior. Pages 359-399, Perspectives in Urban

Entomology, edited by G.W. Frankie and C.S. Koehler. New York: Academic Press, Inc.

Fraser, D.W., J.I. Ward, L. Ajello, and B.D. Plikaytis (1979) Aspergillosis and other systemic mycoses. Journal of the American Medical Association 242(15):1631-1635.

Frieden, B. and A.J. Solomon (1973) The Nation's Housing: 1975-1980. Cambridge, Mass.: Harvard and MIT Joint Center for Urban Studies.

Georghiou, G.P. (1972) Studies on resistance to carbamate and organophosphate insecticides in *Anopheles albimanus*. American Journal of Tropical Medicine and Hygiene 21:797-806.

Goldman, M.I. (1970) The convergence of environmental disruption. Science 170:37-42.

Hall, J. (1976) Subjective Measures of Quality of Life in Britain: 1971-1975, Some Developments and Trends. Social Trends, No. 7. London: Her Majesty's Stationery Office.

Hankiss, E. (1978) Quality of life models (Hungarian experience). Pages 57-96, Indicators of Environmental Quality and Quality of Life. SS/CH/38. Paris: UNESCO.

Healy, G.R., A. Spielman, and N. Gleason (1976) Human babesiosis: Reservoir of infection on Nantucket Island. Science 192:479-480.

Jackson, W.B., P.J. Spear, and C.G. Wright (1971) Resistance of Norway rats to anticoagulant rodenticides confirmed in the United States. Pest Control 39(9):13-14.

Jackson, W.B., E. Brooks, M. Bowerman, and E. Kaukeinen (1975) Anticoagulant resistant in Norway rats as found in U.S. cities. Pest Control 43(4):12-16; 43(5):14-24.

Levenson, H. and G.W. Frankie (In Press) A study of homeowner attitudes and practices toward arthropod pests and pesticides in three U.S. metropolitan areas. Hilgardia.

Maddy, K. (1976) Current considerations on the relative importance of conducting additional studies on hazards of field worker exposure to pesticide residues as compared to studying other occupational safety hazards on the farm. Presented at Pesticide Residue Hazards to Farm Workers Workshop, National Institute of Occupational Safety and Health, Salt Lake City, Utah, February 9-10, 1976. Washington, D.C.: U.S. Department of Health, Education, and Welfare.

Mann, J.B., J.E. Davies, and R.W. Shane (1967) Occupational pesticide exposure and renal tubular dysfunction. Pages 219-226, Acute Glomerulonephritis, edited by J. Metcalf. Boston: Little, Brown & Company, Inc.

McCaull, J. (1976) Discriminatory air pollution. Environment 18(2):26-31.

Milbraith, L.W. (1979) Policy relevant quality of life research. Annals of the American Academy of Political and Social Science 444:32-45.

Moore, R.M., Jr., R.B. Zelmer, J.I. Moulthrop, and R.L. Parker (1977) Surveillance of animal-bite cases in the U.S., 1971-1972. Archives of Environmental Health 32(6):267-270.

National Institute of Mental Health (1972) Pollution: Its Impact on Mental Health, A Literature Survey and Review of Research. National Clearinghouse for Mental Health Information, Report No. 72-9135. Washington, D.C.: U.S. Department of Health, Education, and Welfare.

Nelson, B.C. (1978) Ecology of medically important arthropods in urban environments. Pages 87-124, Perspectives in Urban Entomology, edited by G.W. Frankie and C.S. Koehler. New York: Academic Press, Inc.

Olkowski, W., H. Olkowski, R. van den Bosch, and R. Hom (1976) Ecosystem management: A framework for urban pest control. Bioscience 26:384-389.

Pollard, Z.F., W.H. Jarret, W.S. Hagler, D.S. Allain, and P.M. Schanz (1979) ELISA for diagnosis of ocular toxocariasis. Opthomology 86:743-749.

Pratt, H.D., B.F. Bjornson, and K.S. Littig (1976) Control of Domestic Rats and Mice. HEW Publication No. (CDC) 76-8141. Center for Disease Control. Atlanta: U.S. Department of Health, Education, and Welfare.

Sanborn, G.E., J.E. Selhorst, V.P. Calabrese, and J.R. Taylor (1979) Pseudo cerebri and insecticide intoxication. Neurology 29:122-127.

Surtees, G. (1971) Urbanization and the epidemiology of mosquito-borne disease. Abstracts of Hygiene 46:121-134.

U.S. Congress, Senate (1976) Environmental Protection Agency and the Regulation of Pesticides. Staff Report to the Subcommittee on Administrative Practice and Procedure of the Committee on the Judiciary of the U.S. Senate. December 1976. 94th Congress, 2nd Session.

U.S. Department of Housing and Urban Development (1977) 1976 Statistical Yearbook, Annual Housing Survey. Washington, D.C.: U.S. Department of Housing and Urban Development and U.S. Department of Commerce, Bureau of the Census.

U.S. Environmental Protection Agency (1971) Our Urban Environment and Our Most Endangered People. Task Force on Environmental Problems of the Inner City. Washington, D.C.: U.S. Environmental Protection Agency.

U.S. Public Health Service (1968-1977) Annual Surveillance Reports. Center for Disease Control. Atlanta: U.S. Department of Health, Education, and Welfare.

U.S. Public Health Service (1979a) Reported morbidity and mortality in the United States, 1978. Morbidity and Mortality Weekly Report, Annual Supplement 27(54):3.

U.S. Public Health Service (1979b) Rickettsial Disease Surveillance Report No. 1, 1975-1978. Center for Disease Control. Atlanta: U.S. Department of Health, Education, and Welfare.

Van Arsdol, M.D., Jr., G. Sabagh, and F. Alexander (1964) Reality and the perception of environmental hazards. Journal of Health and Human Behavior 5:144-153.

Walter, T., J. Leavitt, and B. Woody (1979) East New York Housing Plan. New York: New York City Housing Development Administration.

Whorton, D., R.M. Krauss, S. Marshall, and T.H. Milby (1977) Infertility in male pesticide workers. Lancet 2(8051):1259-1261.

Winkler, W.G. (1977) Human deaths induced by dog bites, United States, 1974-1975. Public Health Reports 92(5):425-429.

Winkler, W.G. and K.D. Kappus (1979) Human antirabies treatment in the United States, 1972. Public Health Reports 94(2):166-171.

Woody, B., R.W. Walters, and D.R. Brown (1980) Neighborhoods as a power factor. Society. May/June. (In Press)

World Health Organization (1972) Health Hazards of the Human Environment. Geneva: World Health Organization.

Wright, J.W. (1976) Insecticides in human health. Pages 17-34, The Future for Insecticides: Needs and Prospects, edited by R.L. Metcalf and J.J. McKelvey. New York: John Wiley and Sons, Inc.

2

Major Urban Pests

THE IMPACT OF PESTS ON THE HUMAN POPULATION

"Pests" are a heterogeneous group of animals and plants characterized by the fact that they damage or otherwise adversely affect people or their property. This section focuses on the impact of selected arthropods and includes more general discussions of the impact of vertebrate and plant pests. (Background on a variety of other pests—including weeds, plant pathogens, and wood-destroying pests other than termites—was made available to the Committee. See list of working papers, Appendix B.) Table 2.1 summarizes selected pest-related diseases with current importance in urban areas of the United States.

The root of the word "pestilence" is "pest"—a reminder that over the course of human history the greatest hazard from pests has been the occurrence of life-threatening, epidemic disease. Many of the microorganisms that cause human disease are also parasites of animals that live close to human beings. As Hans Zinsser put it 45 years ago, "There is a lively interchange of parasites between man and the animal world" (Zinsser 1935). Much of this interchange is carried out with the animals that we call "pests."

The magnitude of past epidemics of pest-borne human diseases is staggering. Plague, which originates in rats and other rodents and is transmitted by fleas, carried away one-quarter of the population of Europe—an estimated 25 million people—in three great epidemics in the fourteenth century. The most recent epidemic of plague, which began in

43

TABLE 2.1 Pest-related Diseases with Current Importance in Urban Areas of the United States

Disease	Prevalence	Severity	Pest	Geographic Distribution	Population Affected	Potential for Increased Risk	Prevention or Control
St. Louis encephalitis	>200 cases yearly of proven encephalitis	Severe— ~7% fatal	Culex mosquito-vectors; birds are reservoir for virus	All except New England	Both urban and rural >40 years old	Major epidemics	Surveillance coordinated with vector control
California encephalitis	~100 cases yearly of proven encephalitis	Death uncommon; mild neurologic impairment frequent	Aedes triseriatus mosquitoes—major vectors and reservoirs for virus	Most north central and fewer east and southeast states	Children near wooded areas, including urban and suburban	Not established	Elimination of vector breeding sites
Western equine encephalitis	~50 cases yearly of proven encephalitis	Severe—~2% fatal 15% in outbreaks	Culex tarsalis—major mosquito vector; birds are reservoir for virus	Western and north central states	Mostly rural and small town; young children most severely affected	Often occurs in epidemics	Avoidance of mosquito bites in outbreak; control of adult vectors in outbreak near population centers
Eastern equine encephalomyelitis	< 5 cases yearly of proven encephalitis	Quite severe— 50% fatal	Aedes solligitans are probable vectors for human infection with virus	Eastern seaboard, Texas to New Hampshire	Typically rural–suburban near marsh	Human cases follow bird epizootics	Avoidance of mosquitoes during outbreak; control of vectors near population centers during outbreaks
Rocky Mountain spotted fever	>1,000 cases yearly of proven infection	Moderate to severe—5% fatal-1977	Major vectors for human infection with this ricketsia are ixodid	Most of U.S.; highest in south Atlantic states	Especially children and young adults with outdoor activities		Tick control in residential areas; tick removal; antibiotics for patients

		ticks: *Dermacentor variabilis*, *Dermacentor andersoni*, *Amblyomma americanum*			
Scabies	Mild; occasionally disabling	Poorly documented; ~3% patients visiting dermatologists have scabies	Infestation with itch mite— *Sarcoptes scabiei*	All areas	All; crowding a factor	Individual treatment with acaricide
Pediculosis	Mild	Poorly documented studies of outbreaks in N.Y., Ga., Fla. showed 8% prevalence of head lice on school children, 1973-74	Infestation with head louse— *Pediculus humanus capitis*; crab louse— *Phthirus pubis*; body louse— *Pediculus humanus humanus*	All areas	Especially school children (head lice); crowding a factor	Individual treatment with pediculicide
Arthropod envenomation	>40 reported deaths yearly	Undetermined; probably high	Stinging hymenoptera—especially honeybees, yellowjackets; wasps most important; spiders, scorpions and centipedes less important	All areas	Both urban and rural	Public education; identification and treatment of hypersensitivity

TABLE 2.1 (continued)

Disease	Prevalence	Severity	Pest	Geographic Distribution	Population Affected	Potential for Increased Risk	Prevention or Control
Larva migrans	Poorly documented >5% children have antibodies	Usually mild; leading cause of pediatric retinal disease in southern U.S.	Poorly controlled dogs and cats infected with Toxocara worms	All, but most prevalent in South	Especially children		Prevent fouling of public areas and playgrounds with canine and feline feces
Animal bites	>1,000,000 cases yearly	Usually mild; fatalities uncommon	Dogs and other mammals	All areas	In urban areas, especially children and aged	High in rabies outbreak	
Dog bites	>800,000 cases yearly		Feral and uncontrolled dogs			Trend to larger breeds; rabies risk reduced with vaccination program	Eliminate unrestrained dogs (>90% bites in cities)
Rat bites	~40,000 cases yearly	Rat-bite fever may develop	Rats—*Rattus norvegicus* throughout U.S.; *Rattus rattus* in southern and Pacific states			No rabies risk	Rat control

Bat bites		Risk of rabies high	Bats—many species; some in each U.S. city		Rabies enzootic in bats but epizootics may occur	Eliminate bat colonies from inhabited structures
Systemic fungal infections	Histoplasmosis and cryptococcosis cause 30,000 days of hospitalization yearly	~150 deaths yearly from histoplasmosis and cryptococcosis	Infectious agents develop in accumulations of excreta associated with pigeon and other bird roosts and bat colonies	All areas; histoplasmosis most prevalent in Ohio and Miss. River valleys		Preventive measures for persons removing soil with bird or bat excreta; avoidance of bird and bat harborage sites
Enteric infections	Salmonella and shigella infections common in U.S.; pest role in transmission probably minor at present	Usually mild; fatalities uncommon with good medical care	Can be transmitted via direct or indirect fecal contamination of food and water by rodents, flies, and cockroaches	All; especially those areas with poor sanitation and housing	Breakdown of sewage disposal or general sanitation	Sanitary disposal of wastes; correct preparation of food; control of pest infestations

China in 1894, spread to India by 1898 and caused half a million deaths in that country each year for the next 20 years—a total of more than 10 million deaths (Marks and Beatty 1976). The last major epidemic of typhus, a disease of rats and fleas that also becomes a disease of human beings and lice, occurred in European Russia during and after the Revolution. From 1917 to 1923 there were more than 30 million cases of typhus resulting in 3 million deaths in Russia (Zinsser 1935). In 1793, an epidemic of yellow fever, a disease transmitted to and between human beings by mosquitoes, caused 5,000 deaths in Philadelphia, then a city of 55,000 people (Marks and Beatty 1976). In 1802, yellow fever killed 22,000 of the 25,000 troops that Napoleon sent to Haiti to suppress the revolution of Toussaint L'Ouverture. At present, malaria, another disease transmitted by mosquitoes, infects hundreds of millions of people and each year causes 7.5 million deaths (WHO 1976).

The conditions that set the stage for these pandemics do not now exist in the United States. The black (roof) rat, which lived close to man in the small, crowded, rush-strewn houses of medieval times, has been displaced, for the most part, by the larger, more aggressive brown (Norway) rat, which lives farther from man—in back yards, industrial and commercial buildings, sewers, dumps, or on farms—and finds the more tightly constructed dwellings of today in both Europe and the United States less accessible. In the southern United States and particularly in California, however, the roof rat, supported by luxuriant suburban landscapes and relatively open construction of houses, has become the "bare-tailed squirrel" of the environments and has potential association with native, plague-bearing rodents. The oriental rat flea (which transmitted plague from rats to people) and the human flea (which transmitted it from person to person) are now largely absent from our cities, where most of the fleas today are cat fleas. Human fleas and human body lice diminished markedly when the habits of regular bathing and changing and washing of clothing became customary in the nineteenth century. The settings of human disaster—war, dislocation, poverty, malnutrition, and the presence of other epidemic diseases—in which past epidemics of human typhus have most often occurred are not now often present in our society. Effective quarantine procedures, the suppression of yellow fever in the tropics, and the elimination of human reservoirs of malaria and yellow fever in the United States have greatly diminished the hazard of these two diseases. All this makes it less likely that major new pandemics of the old pest-borne diseases will appear in the United States.

PEST-BORNE EPIDEMIC DISEASE

Nevertheless, there continues to be the possibility that significant outbreaks of pest-borne diseases could occur in American cities and towns. There is a chance for the occurrence of disease wherever people are exposed to insects and other animals that bite, suck blood, sting or otherwise puncture the skin, or deposit their feces, urine, saliva, or other secretions on human food or water or in places where such secretions can be inhaled (Nelson 1978). When human beings live in proximity to animals that harbor infectious agents, there is danger that the agents will be transmitted to human beings, and the presence of potential insect vectors enhances this risk. All these conditions exist and under certain conditions could become widespread in some urban areas of the United States today.

Plague

Endemic foci of plague exist among the small chaparral mammals of the West, some of which live close to human habitation in rural and suburban areas. Peri-domestic harborage of wild rodents and failure to control fleas on dogs and cats have been shown to be risk factors (Mann et al. 1979). In some of these areas, the black rat (now more often called the roof rat) is resurgent. The rat flea is present, and the cat flea (which is capable of transmitting plague) is widely prevalent in human habitations. Cases of human plague have occurred in the western United States. An epizootic of plague among black rats is possible, and the transmission of plague to people and between people could occur (Bahmanyar and Cavanaugh 1976).

Dengue

Dengue is prevalent in the Caribbean area and in parts of Africa and Central and South America. *Aedes aegypti*, the mosquito vector of dengue, is prevalent in many urban areas of the southern United States. Travelers or immigrants entering the country with infectious dengue could provide a source of infection for other people, which could lead to the establishment of dengue in the American population and, perhaps, to the occurrence of epidemics (U.S. PHS 1980a).

Yellow Fever

Jungle yellow fever, for which monkeys serve as a reservoir, is occasionally present near major American and African cities. If the infection was introduced into these cities, it could cause an urban outbreak transmitted by the same mosquitoes that transmit dengue. With increased travel, such an outbreak could lead to introduction of the disease to receptive cities in the United States, although the possibility seems remote because vaccines are available and fewer foci exist.

Mosquito-borne Encephalomyelitis

Mosquito-borne encephalomyelitis is present in this country in several forms, and recognized epidemics of several hundred cases with mortality rates of 5 to 10 percent have occurred (U.S. PHS 1968-1977). The culicine mosquitoes that transmit these diseases, as well as the pigeons, sparrows, starlings, and horses, squirrels and other large mammals that are among its reservoirs are prevalent in and around urban areas. A large population of susceptible people is accessible to the transmitting agents, and it is reasonable to assume that further epidemics could occur.

Malaria

Malaria, which is increasingly resistant to chemotherapeutic agents, now exists in the United States as a disease transmitted from person to person among drug addicts who share the use of subcutaneous and intravenous needles. Infective malarial illnesses are also present in several hundred people from various parts of the world who enter the United States each year, including the increasing numbers of immigrants from Southeast Asia and from Central America. Some of these immigrants have settled in close proximity to each other. Since the anopheline mosquitoes that transmit malaria are present in and around many communities in the United States, especially in the Southeast and in the Gulf states, it seems possible that endemic malaria could become reestablished in this country.

Louse-borne Diseases

Although human body lice are rare in the United States, head lice are common, especially among school children and adolescents. Body lice can transmit typhus and relapsing fever; the ability of head lice to do so under natural conditions remains to be established. There are still foci of murine (rat-borne) typhus in the southeastern United States. Probably more important, there is the possibility that a louse-infested immigrant or traveler who is, in effect, a human reservoir of typhus ("Brill's disease") might come in contact with body-louse-infested susceptibles in this country. The contact could produce a small epidemic of louse-borne typhus similar to those that occurred in courtrooms in Great Britain in the eighteenth century when otherwise healthy but louse-infested people were exposed to prisoners with "jail fever," as typhus was then called.

Rapid Travel by Air

The danger that pest-borne diseases not now prevalent in the United States might be introduced into this country has increased because of rapid transportation of people and goods between far-flung parts of the world. Venezuelan equine encephalitis, which is endemic in Central and South America, was unintentionally introduced into the United States in 1971, and 88 clinical cases were recognized before the outbreak was controlled (Bowen et al. 1976). Other diseases might have more serious consequences. Rift Valley fever is an example; this viral disease affects cattle and people and is transmitted by mosquitoes now present in the United States. The disease was formerly limited to the sub-Saharan regions of Africa, but it has recently spread to Egypt and Sudan. In 1977 there were 18,000 cases and 598 deaths from this disease officially reported by the Egyptian government, but unofficial figures were much higher (Meegan 1979). If it were accidentally introduced into the United States, it could have serious consequences and be difficult to eradicate.

New or Hitherto Unknown Diseases

A hazard of unknown magnitude may exist in the occurrence of variant or hitherto unknown forms of pest-borne diseases. Many such diseases were identified in the United States in the early years of this century, and new ones have continued to be found, e.g., rickettsial pox in the 1940s, babesiosis in the 1960s, and Lyme arthritis in the 1970s. It is not clear whether these are new human diseases or old diseases previously unrecognized.

Babesiosis is a good example (Anderson et al. 1974, Healy et al. 1976). The malaria-like protozoan (*Babesia microti*) that causes this disease is a natural parasite of field mice and other small mammals and is transmitted among rodents by ticks. Although this organism can infect larger wild and domestic animals such as deer and cattle, the disease had not been recognized in humans until it appeared in the 1960s on Nantucket Island as "berry pickers' disease" (Ruebush et al. 1977a, 1977b). Since then several cases have been observed on Nantucket Island. Additional cases have also been observed on other islands of Long Island Sound, on the eastern tip of Long Island itself, and in Georgia. The illness is severe and the symptoms—shaking, chills, fever, severe myalgia, hepatitis, and disorders of blood clotting—are so striking that it is unlikely that the disease could have been previously overlooked, although it might have been mistaken for malaria or some other infectious disease. However, the appearance and spread of babesiosis raises the disturbing question of whether the disease might be another example of an adaptation of a hitherto wild parasite to human hosts. If the parasite of babesiosis should become adapted to transmission by other vectors (such as mosquitoes) that could infect people, major human epidemics might occur. A similar phenomenon is thought to have occurred with the rickettsia of typhus prior to the fifteenth century, when these parasites of rats and fleas became adapted to infecting humans and to being transmitted from person to person by human lice.

The possibility that infectious agents that have hitherto infected only wild animals might become adapted to transmittal by arthropod pests that prey upon human beings makes it difficult to view with equanimity the fact that many people in the United States are now infested with head lice and crab lice or exposed to cat fleas, even though none of these pests is now known to be transmitting serious human disease.

DEATHS, DISEASES, AND INJURIES CAUSED BY PESTS

Directly or indirectly, arthropods and vertebrates that have been defined as "pests" cause an estimated 100 to 300 deaths and perhaps 20,000 cases of disabling disease and injury in the United States each year (U.S. PHS 1968-1977). The actual numbers may be greater than these, because the diseases in question are often not recognized or diagnosed. In addition, pests affect many more people with painful, annoying, disfiguring, or partly disabling conditions resulting from bites or stings or physical reactions to pests and their excreta. Pests also prevent large numbers of people from enjoying their houses, yards, parks, or places of work.

Diseases Transmitted by Blood-sucking and Biting Pests

Blood-sucking and biting arthropods and vertebrates transmit infectious microorganisms directly to human beings from other human beings and animals. These microorganisms include viruses causing encephalitis, rabies, dengue, and yellow fever; rickettsia causing Rocky Mountain spotted fever, rickettsial pox, and typhus; bacteria causing tularemia, plague, and Haverhill fever; and larger one-celled organisms like the parasites of malaria and babesiosis.

The arthropods in urban and suburban areas that transmit disease in this manner include ticks (Rocky Mountain spotted fever, Lyme arthritis, tularemia, babesiosis), mites (rickettsial pox), fleas (plague), and mosquitoes (encephalitis, dengue, yellow fever, malaria). The vertebrates that transmit disease in this manner include rats (Haverhill fever), dogs, skunks, bats, and other carnivores (rabies). Each year these diseases cause from 50 to 150 deaths and probably 10 times this number of disabling illnesses in the United States.

Mosquito-borne Encephalitis St. Louis encephalitis, for which English sparrows and other birds serve as a reservoir, is the leading cause of epidemic encephalitis in the United States. In an average year there are approximately 200 cases, with a 7 percent fatality rate (U.S. PHS 1968-1977). In 1975, there were more than 2,000 laboratory-documented cases with 142 deaths. This form of encephalitis has been reported in all the contiguous United States except South Carolina and the New England states. There is currently no vaccine or specific medical treatment for the disease. Although the classic setting for St. Louis encephalitis is urban, its ecology and appearance vary widely in different areas of the United States and from one outbreak to another.

The vectors of St. Louis encephalitis significant in human infection are the *Culex pipiens* mosquitoes in eastern North America, except for Florida, where *Culex nigripalpus* has been the major vector. In the western states, *Culex tarsalis* has been the major vector; there the disease has been more rural in distribution and has occurred mixed with outbreaks of western equine encephalomyelitis. In the eastern states the classic pattern has been urban, although widespread suburban and rural outbreaks have occurred.

There are approximately 100 cases each year of California encephalitis (which is less often fatal but causes neurological impairment) chiefly in children in urban and suburban areas of the north central states (U.S. PHS 1968-1977). The tree-hole breeding mosquito, *Aedes triseriatus*, is the major vector and squirrels and other rodents serve as the reservoir.

Western equine encephalomyelitis, transmitted largely by *Culex tarsalis* mosquitoes in western states, is more severe, with a 2 to 15 percent fatality rate among the approximately 50 cases that occur annually (U.S. PHS 1968-1977). Young children have been among those most severely affected. Eastern equine encephalomyelitis, which has had a 50 percent mortality rate, is rare except in epidemic years. The five or fewer cases that occur most years are located chiefly in suburban or rural areas. Human cases typically appear in conjunction with epizootics among the birds that are the principal reservoirs.

Tick-borne Rickettsial Diseases   There are approximately 1,000 cases of Rocky Mountain spotted fever in the United States each year, occurring mostly in children and young adults in suburban residential areas, especially in the southeastern states (Hattwick et al. 1973). The mortality rate for the disease was 5 percent in 1977 and has been as high as 20 percent in untreated cases. The number of cases reported annually has increased steadily from about 250 in the early 1960s to 1,063 in 1978 (U.S. PHS 1979b).

Rocky Mountain spotted fever is a rickettsial disease occurring nationwide; it is transmitted from one generation of ticks to another by transovarian infections, and from ticks to animals and people by tick bite. Three species of ticks are of primary importance in the transmission of this disease: the American dog tick, the Rocky Mountain wood tick, and the lone star tick. The American dog tick (*Dermacentor variabilis*) is the chief vector of the disease in eastern and southern states. Larvae and nymphs of the species feed exclusively on small rodents, while adults usually engorge on large or medium-sized mammals, especially dogs. Adult ticks are also found occasionally on cattle, horses, cats, foxes, and humans, but rarely on smaller mammals. The tick is widely distributed east of the Rocky Mountains and also occurs on the Pacific Coast and in parts of northern Idaho and eastern Washington. It is well established in many urban areas; for example, it was collected and identified from 153 outdoor locations in New York City during the 1978 tick season (New York City Department of Health 1978). The Rocky Mountain wood tick (*Dermacentor andersoni*) is an important vector in the West. In its immature stages it attacks small mammals, and its adults feed on larger mammals, including human beings. The lone star tick (*Amblyomma americanum*) is an important vector in parts of the eastern United States and in Texas, Oklahoma, and Arkansas.

Tularemia   A bacterial disease of wild animals—mostly rabbits and rodents—tularemia is transmitted to human beings by tick bite and by contact with infected animals. The ticks include the American dog tick, the lone star tick, and a species found in the Pacific coastal regions (*Dermacentor occidentalis*). Tularemia is widely distributed throughout the United States, but the number of cases reported annually declined from 920 in 1950 to 141 in 1978 (U.S. PHS 1979a). The fatality rate is approximately 5 percent.

Rabies   A viral disease of vertebrates, rabies is transmitted to humans by animal bites, chiefly those of rabid dogs, raccoons, foxes, skunks, and bats. There were only five laboratory-confirmed cases of rabies reported from the United States in 1979. A history of dog bite was established in two cases; the source of infection for the other three was unknown (U.S. PHS 1980b).

The public health problem created by rabies is not so much the occurrence of the disease itself as the potential consequences of a bite from an infected animal. Although bites from rabid domestic animals are rare in U.S. cities, the threat of rabies following bites from dogs is responsible for almost two-thirds of the 30,000 prolonged, painful, expensive, and sometimes dangerous antirabies treatments given in the United States every year (Winkler and Kappus 1979). Although virtually none of the dogs involved is rabid, antirabies vaccinations must be recommended when the biting animal cannot be identified or located.

Bites from animals other than dogs account for only a small proportion of rabies treatments. Nevertheless, bites from bats and wild animals present a threat of rabies infection in every major metropolitan area of the continental United States. Rabid bats have been reported in all the contiguous states. All the common U.S. species have been shown to be infected on occasion. Bat colonies frequently are found in attics and construction voids of older residences, public buildings, and churches.

Bites from domestic rats and mice in the United States are not associated with rabies infection (U.S. PHS 1976). Wider knowledge of this fact might prevent some needless antirabies treatment.

A number of other diseases transmitted by the bites of pests usually cause fewer deaths or disabling illnesses annually, but several of them have the potential for occurring in major epidemics.

Babesiosis A newly recognized disease, babesiosis, caused by a protozoan, has been described above. Ticks (the American dog tick and other ixodid ticks) are the transmitters of this disease.

Lyme Arthritis Ticks are suspected in the transmission of Lyme arthritis, a painful condition that has recently been recognized in the residents of suburban and exurban regions of southern Connecticut, including the town of Lyme from which it received its name. The causative organism has not been identified but is thought to be a virus.

Rickettsial Pox Rickettsial pox is a disease of mice caused by *Rickettsia akari* that is transmitted to human beings by the house mouse mite, *Liponyssoides sanguineus*. Chills, fever, malaise, and a rash that looks like chicken pox are the human symptoms. The fatality rate is less than 1 percent even without specific therapy.

Rickettsial pox is an urban disease that was discovered after an outbreak occurred among adults in a housing development in New York City in the summer of 1946. Cases of it have been noted in many other urban areas of the Northeast and Midwest. In New York City, the number of cases reported annually declined from about 150 in the late 1940s to 2 in 1967 (John Marr, New York City Department of Health, personal communication, 1979). In 1976 and 1978 cases were again reported, and 5 cases appeared in the summer of 1979 (Rafi Al-Hafidh, New York City Department of Health, personal communication, 1979). The reemergence of the disease appears to be related to an increase in the city's mouse population.

Plague   Plague is caused by a bacterium, *Yersinia pestis*. Bubonic plague is characterized by inflammation and swelling of the lymph glands (buboes)—usually those of the armpits and groin—severe toxemia, high fever, prostration, delirium, coma, and in some cases a secondary invasion of the lungs which results in a highly contagious pneumonic disease. Massive septicemia occurs in some cases. Untreated bubonic plague has a fatality rate of 50 to 60 percent (in the fourteenth century apparently the rate was much higher). The pneumonic form is almost invariably fatal.

The classical bubonic flea vector is the oriental rat flea (*Xenopsylla cheopis*). Although not as abundant as other fleas in the United States, it is nevertheless well established throughout the country and is in fact one of the most abundant rat fleas in the southern states and in southern California. People can also acquire the infection by handling the tissues or pus of infected animals. When the pneumonic form of the plague occurs, it is spread through the air by droplets of sputum exhaled by infected people.

Sylvatic (wild rodent) plague exists in many parts of the world, including the western United States, where it occurs in ground squirrels, chipmunks, woodrats, deer mice, prairie dogs, and other native rodents. Occasionally, epizootics occur among these rodents, and large numbers of them die as a result. Infection could be transferred to domestic rats in urban or suburban areas where rats and native rodents comingle. Approximately 15 cases of sylvatic plague in human beings are reported in the United States each year, with a 10 percent mortality rate (U.S. PHS 1979a). Most of the cases have occurred in western states among people who have come into contact with wild rodents or their fleas. Rat-borne human plague has not been seen in the United States since 1925, but it remains a real and apparently increasing threat wherever commensal rats come into contact with enzootic or epizootic plague in wild rodents in suburban areas (Mann et al. 1979).

The human flea (*Pulex irritans*) has been considered a vector for the spread of plague in the past. This flea is not as abundant in human populations as it used to be, but it is still found through most of the United

States and is still among the most common of fleas in homes along the Pacific coast. *Pulex irritans* attacks a wide variety of hosts other than humans, including wild and domestic animals. It is not normally a carrier of plague, but it remains a potential transmitter.

Ctenocephalides felis, the cat flea, is the most common flea in and around human dwellings in the United States. It is found throughout most of the country but is somewhat less common in the Rocky Mountain states. It prefers dogs, cats, or human beings as its host, but it attacks a wide variety of other mammals, including rats, and is most prevalent during the last part of summer. In the laboratory, the cat flea has been shown to transmit plague (Pratt and Wiseman 1962). It is an inefficient vector compared with the oriental rat flea, and it has never been identified as the primary vector in a human epidemic. Although it must be recognized that cats more frequently, and dogs occasionally, succumb to plague when there is an epizootic among rodents, the most likely mode of infection is thought to be by ingestion of infested rodent prey. The potential of cat and dog fleas to transmit plague to humans remains, although it has not been demonstrated.

Ctenocephalides canis, the dog flea, is similar to the cat flea in its biology and habits. Although less common than the cat flea, the dog flea has been found in all parts of the country with the exception of the Rocky Mountain and intermountain regions.

Dengue In 1978 there were 89 cases of laboratory-diagnosed dengue imported into the continental United States by travelers from the Caribbean and Central America (Kappus et al. 1979). The large number of recent immigrants from Mexico, Puerto Rico, and Central America, where the disease is widely endemic, enhances the probability that dengue will also occur in this country. Recent outbreaks of this disease in the Western Hemisphere have been mild. Although more than 200,000 cases were estimated to have occurred in Puerto Rico in 1977, there were no deaths. Dengue is transmitted from person to person by *Aedes aegypti*, a highly adaptable urban mosquito that breeds almost exclusively in discarded containers in and around human dwellings. The species is found throughout most tropical and subtropical regions of the world. In this country, it is widely distributed and fairly common in the southern and Gulf Coast states. Although populations of *Aedes aegypti* have been found as far north as New York and Illinois, it appears that they are introduced during warm seasons from more temperate areas, and that they multiply during the summer months and die in the winter. Recent surveys by entomologists from the U.S. Public Health Service's Center for Disease Control revealed the presence of this species in 29 of 30 cities surveyed in

10 southern states (Don Eliason, Bureau of Tropical Diseases, Center for Disease Control, personal communication, 1979).

Yellow Fever *Aedes aegypti* is also the vector of yellow fever, a severe viral disease with a high mortality rate, endemic in Africa and Central and South America. The last epidemic of yellow fever in the United States occurred in 1906; the last imported case was identified in 1924.

Both dengue and yellow fever are of concern because of the wide distribution of potential vectors in the United States among highly susceptible human populations. Outbreaks of yellow fever also occur in the populations of tropical countries but have been confined largely to jungle yellow fever in rural regions of the Americas, although occasional outbreaks of urban yellow fever are reported from Africa. The probability that travelers with infectious yellow fever will enter the United States is small but not negligible. Air travel brings large numbers of people to the United States from all parts of the world, and it will be only a matter of time before yellow fever is again found in this country.

Malaria Mosquito-transmitted malaria is not now endemic in the United States, but the two major vector species of anopheline mosquitoes, *Anopheles quadrimaculatus* in the eastern states and *Anopheles freeborni* west of the Rockies, are highly prevalent in many fresh-water habitats. Both are vectors of the various protozoan parasites of malaria. There were 616 imported cases of malaria and 6 deaths reported in the United States in 1978 (U.S. PHS 1979c). Many of the cases occurred among recent immigrants who settled in areas where the potential mosquito vectors of malaria were present.

Rat-bite Fever Rat-bite fever (also called Haverhill fever) is a serious, generalized infection caused by *Streptobacillus moniliformis* and transmitted by the bites of rats. The number of cases that occurs annually is not known. Epidemics have been reported in the past, however, and one survey of rat bites in Baltimore indicated that rat-bite fever occurred in 11 percent of 87 cases (Brooks 1973).

Tetanus Cases of tetanus occur from time to time as a result of either dog bites or rat bites.

Diseases Transmitted by the Feces and Urine of Pests

Some pests, both arthropod and vertebrate, transmit disease by depositing their urine or feces on people's skin, on food, in water used for drinking or bathing, and on floors, counters, clothing, grassy areas, and other places where people may come into contact with them. Other pests deposit their urine or feces in places where it dries and becomes dust containing infectious agents, which people then inhale. Among these agents are the rickettsia of typhus, the bacteria of enteric infections, the spirochetes of leptospirosis, the larvae of the dog and cat tapeworm, and a variety of one-celled fungal organisms.

The pests that can cause disease in this manner include lice and fleas, which transmit typhus by depositing their feces or stomach contents on human skin; rodents, whose feces may contain *Salmonella* or other enteric organisms; cockroaches, whose feces may transmit several kinds of enteric diseases; rodents, whose urine in drinking or bathing water may transmit leptospirosis; dogs and cats, whose feces in houses, parks, and playgrounds may transmit tapeworms; and pigeons, starlings, and sparrows, whose dried fecal accumulations support fungal growth and, when dried or disturbed, may be inhaled.

The number of fatal illnesses caused in human beings each year in this manner by fungal infections alone is estimated to exceed 100 (Fraser et al. 1979). Reliable estimates of the number of fatal or disabling illnesses from the other infections are not available.

Typhus Typhus is transmitted to humans primarily by the feces of lice and fleas deposited on human skin and then rubbed into the skin when the person scratches the bite of the arthropod. Sustained in nature by rat-flea interaction, with the roof rat and Norway, or brown, rat serving as the reservoir, typhus is a rickettsial infection of man and domestic rodents that is worldwide in distribution.

The epidemic form of typhus is a rickettsial infection caused by *Rickettsia prowazeki*. It is spread from person to person by human lice. The disease may be severe; in times of epidemic, or in the absence of specific therapy, the fatality rate varies from 10 to 40 percent (Beneson 1975). The last outbreak of louse-borne typhus in the United States occurred in 1921.

The murine (derived from murid rodents) form of typhus is caused by *Rickettsia typhi* transmitted to humans by infested rat fleas. In general, this form of typhus is somewhat milder than epidemic louse-borne typhus. Mortality for all ages is about 2 percent, but is higher among the elderly (Beneson 1975).

Murine typhus in the United States reached a peak of morbidity in 1944, when 5,400 cases were reported. The incidence of murine typhus decreased rapidly after vigorous rodent and flea control measures were initiated in the southern states. In the United States, about 90 percent of the human cases occur in the southern states and southern California. Fewer than 100 human cases are now reported each year (Pratt and Wiseman 1962, U.S. PHS 1979a).

In the early years of this century it was found that people who had recovered from clinical typhus could continue to be subclinical carriers of the disease, and that it could recur later in life (Brill-Zinsser disease), becoming a source of infection to other people.

The louse that transmits epidemic typhus is the human body louse, *Pediculus humanus humanus*. Body lice spend their lives in clothing, depositing their eggs in its seams and creases and, occasionally, on the hairs of the body. Head lice, *Pediculus humanus capitis*, are usually found in the hair on the head and on the scalp. Females glue their eggs to the hairs of their host.

Lice are transmitted from one infested person to another by direct contact, and indirectly by contact with personal belongings, especially clothing, bedding, and head gear. At present, body lice are not found frequently in the United States. When they are found they occur mostly among vagrants and among chronically ill and inadequately cared-for people in institutions. Head lice, however, are increasingly prevalent among school children and adolescents throughout the country.

Enteric Infections from Rodents Both mice and rats are frequently infected with *Salmonella*, and they are able to cause epidemics of *Salmonella* food poisoning transmitted through their feces deposited on human food. Food poisoning in humans is usually acute and transiently severe but not often fatal.

Lymphocytic Choriomeningitis A viral disease, particularly of mice, lymphocytic choriomeningitis is most commonly transmitted to humans by contact with urine of infected mice but can be transmitted to humans through contaminated food. Mice that are infected may recover but continue to carry the viral organism as long as they live. They can pass the virus to their offspring or shed it in urine, feces, or nasal secretions.

Enteric Diseases Associated with Cockroaches Domestic cockroaches— cosmopolitan insects found in most places where people live and work— are probably the most common pest in urban communities. Fifty-five species of cockroaches are known to live in the United States, although only a few infest dwellings, institutions, and food-handling establishments.

The health problems associated with domiciliary cockroaches in urban environments have been well-documented (Roth and Willis 1957, 1960; Cornwell 1968). Apart from the fact that a major cockroach infestation implies domestic uncleanliness and embarrasses individuals regardless of their socioeconomic status, cockroaches consume human foodstuffs and contaminate them with salivary secretions and excrement. The secretions impart a persistent, fetid odor to materials that they contact.

Several different types of pathogenic organisms have been recovered from cockroaches and their feces. Cockroaches have been found to harbor 3 strains of poliomyelitis virus, about 40 species of pathogenic bacteria, and the eggs of 7 different species of pathogenic helminths (Roth and Willis 1960). There is supporting evidence that cockroaches may be involved in outbreaks of infectious hepatitis and various forms of food poisoning, dysentery, enteric fever, and gastroenteritis in urban areas, but their exact role in outbreaks of human disease is not fully understood (Roth and Willis 1960, Cornwell 1968, Rueger and Olson 1969).

Other disease-producing organisms can be carried by these insects (Roth and Willis 1957), but in general cockroaches are not associated with widespread contagion or outbreaks of disease (Guthrie and Tindall 1968). Cockroaches also are medically significant in that they may induce allergic reactions in human beings (Bernton and Brown 1964, 1970a, 1970b) or seriously contaminate sterile medical equipment in hospitals (Alcamo and Frishman 1980).

Although cockroaches are widely disliked and strongly suspected of being transmitters of serious disease, the evidence that they are actually involved in transmission is scanty.

Leptospirosis A severe and often fatal infection, leptospirosis can be transmitted to human beings by rats, dogs, or other mammals that deposit urine containing leptospires in water used for drinking or bathing, or it can be transmitted by direct or indirect contact with infected animals. Actual incidence is not known because the disease is variable in its course and is probably not fully recognized or reported. Nonetheless, from 5 to 10 deaths have been reported in the United States annually since 1971 (U.S. PHS 1979a). One report from St. Louis documenting the sewer rat to dog to child transmission may be a far more common mode than is suspected (Feigin et al. 1973).

Diseases Associated with Bird Feces Several human diseases are caused by one-celled organisms that are primarily yeast or fungi. The organisms are cultured naturally in the excreta of birds and can be transmitted to humans when they are dried and/or disturbed.

Aspergillosis from exposure to the spores of *Aspergillus fumigatus* has been reported to occur in people who feed pigeons. There are fairly frequent reports of cases of histoplasmosis caused by the inhalation of

Histoplasma capsulatum from dried bird feces. Cases of meningitis due to *Cryptoccus neoformans* have been reported as well.

Pigeons, starlings, and blackbirds are primarily implicated in the transmission of these diseases in urban areas. Pigeons roost on buildings and drop their feces on window ledges and air conditioners from which the feces can be dispersed into homes. Starlings and blackbirds often roost in large numbers in trees, dropping their excreta in parks and roadways. People who inhale dust from bird feces are especially at risk.

Contact with nasal secretions or excreta of birds, especially pet parakeets and pigeons, is also a typical association in the 50 to 150 ornithosis (psittacosis) infections reported in recent years (U.S. PHS 1979a).

Helminth Infestation   The tapeworms that ordinarily infest dogs, cats, and rats are occasionally transmitted to humans through animal feces deposited on human food or accidentally ingested. Some fleas are said to be capable of transmitting the tapeworm through their feces as well.

Dog feces also often contain the ova of *Toxocara canis* that in larval form invade human tissues, including the retina of the eye, and are a leading cause of retinal disease in children in the southern United States. In some areas 10 to 23 percent of the soil samples from parks and playgrounds contain *Toxocara* ova deposited there in dog feces (Schantz and Glickman 1979).

Enteric Diseases Transmitted Passively by Pests

Some pests, notably flies and cockroaches, can transport infectious agents—especially the bacteria and viruses of enteric disease—passively, by carrying the agents on their feet as they move from sewers or animal or human feces to human food. Cockroaches living in sewer systems can act as "elevators" of pathogens as they move through water traps and disperse via other routes into living and working areas as a result of population pressures or sewer environment disruptions (Jackson and Meier 1955, 1961). The magnitude of human illness produced in this manner among urban residents is not known.

Pests as Direct Causes of Human Disease, Injury, or Death

Some pests, through their bites or stings, are direct causes of human disease, injury, or death.

Dog and Cat Bites   In 1972, when data on reportable diseases were available from 20 states, animal bites ranked fourth among causes of such conditions in the United States (Moore et al. 1977). Each year more than 1 million Americans sustain bites of sufficient severity to warrant medical attention. In more than 800,000 cases, the biting animal is a dog; in more than 95,000 cases, it is a cat. In urban centers, the percentage of bites attributed to dogs is approximately 90 percent. Studies of the epidemiology of dog bites in cities have shown that most attacks are by free-ranging or ownerless pets.

Typically, the person bitten is a child (40 percent) or a teenager (28 percent), and the bite is on the extremities (74 percent) (Moore et al. 1977). Most animal bites are painful and are frequently followed by infection. If the biting animal cannot be found or its health status determined, it is often necessary to administer rabies vaccine. The severity of dog bites varies widely, and a study of 11 fatal attacks in the United States during 1974 and 1975 concluded that the cases reported represented only a portion of the actual number that occurred (Winkler 1977).

Bites of Rats and Other Animals   Each year more than 43,000 people in the United States are estimated to be bitten by rats and mice. About 30,000 bites by other animals, including wild animals, are also estimated (Moore et al. 1977).

Studies in several communities indicate that approximately two-thirds of rat-bite victims are children under 10 years of age (Moore et al. 1977). Adults who are attacked by rats are usually helpless or debilitated people, including vagrants, alcoholics, and the aged. Most of the bites are on the hands or the feet, but serious and disfiguring attacks on the heads and faces of infants can occur when rats are attracted to those areas by food. Occasionally, rat bites result in the death of infants or debilitated adults (Pratt et al. 1976). Rat bites may become infected and, as reported above, result in rat-bite fever.

Stings of Venomous Arthropods   A number of arthropods cause painful and sometimes fatal injuries to people when their stings or bites inject venom. The most common of these are bees (including honeybees and bumblebees) and wasps (yellowjackets, hornets, and several other species of wasps); some centipedes (of the genus *Scolopendra*); some scorpions; and some species of spiders, including the brown recluse and black widow spiders.

About 40 deaths and many times this number of painful and disabling injuries result from stings of arthropods in any year (Barnard 1973). Most of the deaths appear to be the result of unusually severe human allergic reactions to the venom of the arthropod.

Tick-bite Paralysis   In North America the bite of the wood tick, *Dermacentor andersoni*, or of the American dog tick, *Dermacentor variabilis*, may lead to a progressive, ascending motor weakness and paralysis believed to be caused by a neurotoxic substance injected by the engorging female tick. The symptoms are most frequently observed in children and usually disappear promptly when the tick is removed. The number of cases of tick-bite paralysis that occur in this country each year is not known.

Scabies   Scabies ("the itch") is a disease caused by human infestation with the itch mite (*Sarcoptes scabiei*). The mite is an ectoparasite of man that spends its entire life cycle on the human host. Females burrow into tunnels in the upper epidermis of human skin to lay eggs that hatch in 3 to 5 days. After going through nymphal and larval stages in about a week, the mites become adults and live for a month or longer. They make burrows most commonly between the fingers, and around the wrists and elbows, male genitalia, buttocks, and axillae. The lesions resulting from reaction to the infestation produce a typical pattern on the body and extremities. In most cases infestation with the itch mite causes intense irritation. Scratching can result in secondary infection of the excoriations. Scabies is usually acquired during direct and extended contact with another infested person, often in a family context.

The itch affects people of all social strata. Children are often infected. The actual extent of infestation in the United States is not known. According to a survey of dermatologists, however, scabies accounts for an average of approximately 3 percent of their practice (Shaw and Juranek 1976).

Bites and Stings of Other Arthropods   A number of arthropod pests, among them fire ants, chiggers, and some biting flies, cause bites and stings that are notably painful. Reactions are rarely fatal but may be extremely uncomfortable. The stings and bites of these insects, like those of mosquitoes and fleas, may become infected as a result of scratching.

Allergic Reactions

Allergic reactions to pests and their excreta are important causes of death and disease in the United States. The injurious effects of the bites and stings of arthropod pests are often complicated and enhanced by allergic reaction to the protein of the pest or to its saliva or excreta. Major allergic reactions to the stings of wasps (including hornets and yellowjackets) and bees are the cause of most of the fatal reactions. There may be 20 or more acute but nonfatal allergic reactions to insect stings for each fatal one, and the number of deaths as a result of allergic reaction to insect stings is about 40 per year (U.S. DHEW 1979). These pests occur in all parts of the United States during the spring, summer, and early fall months.

Induced sensitivity of the human host also contributes significantly to the discomfort caused by the bites of fleas, chiggers, mosquitoes, and mites. Sensitivity to the protein of house mites, which is inhaled as dust, appears to be a significant cause of human asthma and rhinitis.

Several million people each year are probably affected in one way or another by allergic reaction to animal and arthropod pests and pest products.

Plant Pests and Human Disease Although some plants contain poisonous substances and can cause death or serious illness if eaten, the vast majority of adverse human reactions to plants is the result of allergic hypersensitivity. The common allergic reactions take three major forms: reactions involving the upper airways (allergic rhinitis, or "hay fever"), reactions involving the lower airways (asthma), and reactions involving the skin (atopic and allergic dermatitis).

The most frequent natural allergens are protein products of plants and animals, but the capacity of simple chemicals in drugs to become conjugated with or attached to proteins also makes them frequent allergens in the industrial or medical setting.

The extrinsic allergens that most often precipitate rhinitis are inhaled particles about 50 microns in size. These include pollens and dusts of many kinds as well as animal danders. The extrinsic allergens that initiate asthma are most often particles of smaller size but of similar origin. Naturally occurring extrinsic allergens that cause allergic dermatitis are most often plant or animal products that come into contact with the skin.

The allergens are of great medical significance. An estimated 35 million Americans have allergic diseases. Of these, 8.9 million have asthma, 14.7 million (including 5 million children) have allergic rhinitis, and 11.8 million have skin allergies and allergies of other kinds. An estimated 2,000 to 4,000 deaths due primarily to asthma occur each year (U.S. DHEW 1979). Rhinitis is one of the most common causes of short-term disability associated with absence from school or work.

Most allergies represent hypersensitivity to common plants and animals. With few exceptions, plants are not pathological agents that "cause" human disease in the sense that viruses, bacteria, or animal parasites do.

Although allergic illness is more usually the result of an abnormality of the person than a peculiarity of the plant, preventive measures are directed at both the human sufferer and the plant offenders. Because sufferers in general are allergic to so many common plant and animal substances, the elimination of any one of these from the environment would probably not greatly reduce the burden of disease in the general population.

Some plants, like ragweed, are notorious causes of rhinitis in urban areas, and others, like poison ivy, poison oak, and poison sumac secrete highly allergenic substances that cause allergic dermatitis. There are undoubtedly circumstances in which it can be helpful to eliminate ragweed from vacant lots in cities or poison ivy from suburban gardens, but it is always desirable to help the individual sufferer avoid the plants and animals to which he or she is susceptible.

PAINFUL AND ANNOYING ASSAULTS

The painful and annoying assaults made by vertebrate and arthropod pests—biting, stinging, and burrowing—are the most widespread adverse effect they have upon people.

Reactions to Bites and Stings

Bites of fleas, itch mites, bedbugs, mosquitoes, chiggers, and similar pests can significantly degrade the quality of life of otherwise healthy people. A person bitten by such an insect may develop a sensitivity to the salivary products injected with the bite, and subsequent bites can produce greater inflammation and pain, and more itching. Continuation of the bites may enhance the sensitivity, causing repeated scratching that can lead to dermatitis in some of the victims. When infestations become severe enough, people may abandon otherwise useful buildings or avoid outdoor areas they would like to use. The number of people frequently exposed to attacks by one or another of these pests is probably in the millions.

Human Infestation

Itch mites, head lice, pubic lice, body lice, and fleas infest humans directly. Quite aside from the bites associated with the infestation, simply to be infested is itself an annoyance. Scabies ("the itch"), caused by the human itch mite, is one of the most annoying infestations that humans experience. Although human infestation with fleas and body lice is not as common in the United States as in certain other countries, it still probably affects tens of thousands of Americans. Infestations with itch mites and head lice are even more common, and scabies is widespread among school children in the United States. The number of people infested in 1976 with head and pubic lice has been estimated at 3 to 5 million (Elzweig and Frishman 1977).

An equal number is infested with the crab louse (*Pthirus pubis*), which spreads directly from person to person through intimate body contact. It most commonly infests the hair of the pubic and peri-anal regions, but it may also infest the hairy regions of the chest or the armpits and, sometimes, the eyelashes, eyebrows, and beard, although it does not commonly invade head hair. Eggs are glued to hairs in the infected regions, and the louse spends its life on the body of its human host. The pubic louse does not appear to transmit any microbial disease, but it is a major nuisance in itself.

EFFECTS OF URBAN HABITAT ON EXPOSURE

People in modern societies are so mobile that pest-borne diseases cannot be considered limited to urban, rural, or even tropical environments. Often a pest-borne disease is contracted by an urban dweller in a rural setting or in another country, and the traveler returns to an urban environment before illness shows itself. Similarly, many arthropod and vertebrate vectors of disease whose usual habitat is a rural area may penetrate into urban areas, causing illness. Sometimes these pests are transported to urban regions by trucks, vans, or airplanes. Nevertheless, the various habitats of common urban pests can result in their having different impacts on different parts of the urban population.

COMMON PESTS IN SUBURBAN AREAS AND ON THE FRINGES OF URBAN AREAS

Mammalian Reservoirs of Disease

Mammals that are reservoirs of Rocky Mountain spotted fever, tularemia, sylvatic plague, and rabies are found mostly in natural areas, on the fringes of urban areas, and in the suburbs. These animals include woodchucks, rabbits, opossums, skunks, raccoons, ground squirrels, chipmunks, field mice, woodrats, deer mice, marmots, prairie dogs, foxes, and deer. Ticks that transmit rickettsial disease to human beings are located, for the most part, in the same areas. People who live on the fringes of urban areas or in the suburbs run the risk of being exposed to this group of disease carriers.

Bats, recognized as reservoirs of rabies, may be found in rural, suburban, and urban settings, often colonizing human dwellings.

Because some suburban developments in the Southwest and West have spread into areas where sylvatic plague is prevalent, and because the roof rat has spread into some of these developments, people in those parts of the country are among those who are at increasing risk of contracting plague.

Venomous Arthropods

Since bees, hornets, and wasps are found more frequently in suburban areas than in the central city, suburban dwellers are more likely to encounter venomous insects than are inner-city residents. Similarly, heavy infestations of chiggers and fire ants are more likely to occur on the fringes of urban areas, especially in the South and Southeast.

PESTS THAT FREQUENT BOTH RURAL AND URBAN AREAS

Mosquitoes

Heavy infestations of those mosquitoes that transmit malaria and those that are mere nuisances are likely to be encountered in those areas of cities or towns near wetlands and woodlands. On the other hand, mosquitoes that can transmit dengue and yellow fever may be found in central urban areas as well as in suburbs because they can breed in water accumulated in catch basins or similar places.

Vectors of Encephalomyelitis

Although farm animals and small mammals that are reservoirs for some forms of mosquito-borne equine encephalitis are nearly always found in rural areas or on the fringes of urban areas, birds that are reservoirs of these diseases and culcine mosquitoes that transmit them can readily penetrate into central urban areas. Birds, such as house sparrows, can and do propagate in urban areas and are often the source of the viruses that causes the disease there.

PESTS MOST COMMONLY FOUND IN AND AROUND HUMAN DWELLINGS

Some of the most common urban pests are commensals—those that live close to people, inhabit the structures that they build, and eat human food, garbage, or excreta.

Rats

The brown or Norway rat is found most frequently near commercial and residential places where garbage and solid waste accumulate and where food is prepared or stored. It often inhabits poorly maintained houses as well as commercial buildings, warehouses, and dumps where food is plentiful. The roof rat, with its greater climbing ability, may live close to people and can be found in urban and suburban dwellings.

Mice

The house mouse inhabits even the most modern buildings of all cities if food is available. They may enter such structures during construction as a result of poor sanitation practices. The closing of apartment house incinerators in New York City several years ago (an action carried out to reduce air pollution) led to the compacting and storage of solid wastes on the premises of many apartment houses. This in turn has led to a significant increase in the mouse population of the city and, probably, to the recurrence of cases of rickettsial pox in humans. In the suburbs, field mice may move into human dwellings in the fall as the weather becomes colder.

Pigeons, House Sparrows, and Starlings

Pigeons, house sparrows, and starlings live in cities and towns, finding their food in city streets and yards. The habits of these birds, with regard to their droppings, were discussed above. Because of the tendency of both pigeons and starlings to roost close together in large numbers, some buildings or park lands may become thickly covered with bird feces.

Cats and Dogs

Most cats and dogs are pets, and while many are kept under leash, large numbers are allowed to run free by their owners and many others are ownerless. Cats and dogs that run free may frequent basements or abandoned buildings, thus accounting for large accumulations of fleas in these areas. When there are large populations of dogs in cities and towns, the parks and streets may become littered with their feces.

People who live close to dogs and cats or who keep these animals in their homes are more likely to encounter the fleas and ticks that are ectoparasites of these animals. Unmanaged pet and feral dogs constitute one of the most serious urban public health problems.

Flies and Cockroaches

Houseflies and cockroaches are found wherever people live and store their food, most commonly in places where sanitation is poor and solid waste accumulates. Cockroaches are especially plentiful in older kitchens and wooden buildings that provide many cracks and crevices in which food can accumulate and many openings that serve as access to harborage and as passageways for the pests; however, no building containing people, food scraps, and dark areas in which roaches can propagate is entirely free of them.

Human Ectoparasites

Some common pests are human ectoparasites. The head louse, the body louse, the crab louse, the human flea, the itch mite, and the bedbug all fall into this category. They live in the hair, clothing, and bedding of humans. People who wear their hair short and who wash their bodies, their clothing, and their bedding regularly are less likely to suffer sustained infestations of these pests, though they may encounter them from time to time.

EFFECTS OF PESTS ON THE POOR

For a variety of reasons the diseases transmitted by commensal pests and human ectoparasites are most likely to be encountered by those Americans who have the lowest incomes, whether they live in rural or urban areas. In urban areas, poor people often live in old, run-down buildings that provide ready access to pests in areas where sanitation is poor, where solid waste accumulates in dwellings, hallways, alleys, streets, and back yards, and where lack of maintenance and policing allow pests to proliferate without hindrance. The dwellings of low-income people are more likely to be located near commercial buildings or dumps, or in low-lying or poorly drained areas. Facilities for personal bathing and for washing clothing are likely to be inadequate. The close association of some low-income people with others who are infested, their inability to obtain clean bedding or furniture when their own becomes infested, and their inability to move out of infested buildings often mean that even if they do get rid of infestations, their bodies and homes promptly become re-infested. Although conditions such as those described here are commonly thought of as attributes of old multifamily dwellings in central areas of large cities, they are also present in low-income areas on the fringes of cities and towns and in areas where people live in run-down and unsanitary dwellings on separate plots.

It should also be noted that people in low-income groups are more likely to be exposed to pest-borne diseases because they are more likely to have unskilled jobs in slaughterhouses, warehouses, garbage collection facilities, food-processing plants, demolition sites, and other areas likely to bring them into contact with rats, mice, fleas, spiders, bats, or accumulations of animal excreta.

Some of the most damaging exposures to pests occur among the ill, debilitated, aged, or otherwise helpless poor. Vagrants, often alcoholic or mentally ill, may neglect to wash or change clothes, and the aged and debilitated poor who cannot do so are the humans most likely to be infested with body lice. It is also these people and infants, left untended in infested dwellings, who are most likely to be bitten by rats.

PEST RESISTANCE AND RESURGENCE OF PEST- AND VECTOR-BORNE DISEASES

The rapid development of pest resistance to a variety of chemicals has been the major factor that prompted the development and growth of alternative pest management strategies, especially integrated pest management. The phenomenon of pest resistance has been observed in pests of public health importance and could have been anticipated as early as 1947 when the common housefly (*Musca domestica*) and the mosquito (*Culex fatigens*) were first reported to be resistant to DDT in Italy. The development of resistance to insecticides hampers efforts to maintain medically important pests at acceptable population levels. In fact, the development of resistance is a major factor in the increase of malaria in several parts of the world. In agriculture, the initial impact of pest resistance is the increased dosages (and consequent increased cost) required to achieve satisfactory control, but the initial result in anti-vector programs is the resumption of transmission of disease. The mosquito vectors of malaria have demonstrated resistance to DDT, to the cyclodiene derivatives, and to the organophosphate and carbamate insecticides, and the mosquito *Anopheles albimanus*, which is the malaria vector in Central America, is resistant to all four insecticidal groups (Georghiou 1972, Garcia and Najera-Morrondo 1973).

Although the mechanism of acquired resistance is not fully understood, one contributing factor appears to be the selection pressures applied by pest management strategies in agricultural and other pest management programs. For example, in India, pest resistance has occurred in areas where wells have been treated with larvicides and where intensive DDT spraying had occurred in homes for a number of years, but no DDT was used for agricultural purposes. In other parts of the world, malaria eradication research programs have shown that "areas where the specific insecticides are widely used in agriculture generally overlap the areas where the malaria vector is resistant to those insecticides" (Garcia and Najera-Morrondo 1973).

Both physiological and behavioral resistance has been described in the malaria program in Central America. The latter is seen especially in malaria control, which is effected by spraying inside houses; when the vector comes into contact with the insecticides to which it has become sensitized, it flies out of the house, thus replacing the interior feeding habitat with a much more unmanageable outdoor habitat. Behavioral resistance illustrates a phenomenon in which a pest management technique has contributed to vector modification that has facilitated the spread of malaria. Thomas D. Mulhern, former Executive Director of the American Mosquito Control Association, has stated: "More than 83,000 malaria cases were reported in El Salvador in 1976, and high resistance to house sprays has developed in *Anopheles albimanus*, the primary factor" (Neilsen 1979). He also stated that "the greatest resistance in El Salvador resulted from a combination of household spraying and continued exposure of *A. albimanus* to DDT and other chemicals used in controlling agricultural insect pests." Thus, chemicals being used both in agricultural and urban ecosystems can promote insect resistance and lead to resurgence of diseases. Recent increases in dengue fever and pediculosis are examples of phenomena that may be due to pest resistance.

Resistance has been noted in other cases as well. Increases in human head lice have been reported from all areas of the world. In parts of France and the USSR increases of 50 to 60 percent have been observed (Lamizana and Mouchet 1976, Palika et al. 1971). A similar situation has been noted in Chile where increases of 17.3 percent in males and 22.5 percent in females have been observed (Scheone et al. 1973). In the United States in 1976, it was suggested, on the basis of sales of pediculicides such as gamma benzene hexachloride (lindane), that some 6 million cases existed (Anonymous 1976). In recent years, cases of resistance of head lice to the widely used and efficient organochlorines have been reported in Canada, Denmark, England, France, Hungary, the Netherlands, South Africa, and the United States (WHO 1977, Blommers and van Lennep 1978).

It is likely that resistance to chemicals used in urban pest management will continue. As early as 1962, the dangers of complete reliance on chemical control were recognized by the Deputy Director of the Malaria Eradication Unit in Geneva who stated "It is now generally accepted that resistance to insecticides by all disease vectors is inevitable" (Pal 1962).

Additional reviews of resistance in rodents and mosquitoes are presented in Chapter 3.

SURVEY OF PESTS

EXTENSION SERVICE SURVEY

During the summer of 1979, extension services in all the land grant institutions in the United States were asked to name 10 major indoor and 10 major outdoor urban pests or pest groups in their states. Where applicable, information was also requested on pests of importance to public health.

All the institutions responded to the request. The state lists were grouped according to the nine established regions of the Center for Disease Control (CDC) (Figure 2.1), and the lists were then combined to show the major pests in each region; pests reported by fewer than one-quarter of the states in each region were listed separately from the major pests. No attempt was made, however, to list the major pests in order of importance. Most of the extension services indicated that their lists were based on general impressions rather than on a comprehensive study. None of the states indicated that their pests are notable for their public health importance. A complete list of the pests in each CDC region is presented in Tables D.1 through D.9 (see Appendix D).

Tables D.1 through D.9 do not allow convenient comparative examination of pests among the regions, but Tables 2.2 and 2.3 show the major pests regrouped into larger categories and indicate in a general way which pests or pest groups cause concern in a majority of the nine regions.

Overall, Tables 2.2, 2.3, and D.1 through D.9 provide an impressionistic tally of major pest distributions. They also indicate that many states and regions have distinct localized pest problems, as revealed in the lists of "other reported pests" in Tables D.1 through D.9.

It could be argued that the lists present a biased view of the pests in given states. This might be so because (a) only certain segments of the population seek information from university extension services, and (b) people at a distance from land grant institutions are less likely to use extension services as sources of information (Frankie and Levenson 1978, Levenson and Frankie, in press). In order to assess the accuracy of the extension service lists, the lists were then compared with findings from other studies.

COMPARISON OF LISTS FROM CALIFORNIA

Two hundred homeowners in Berkeley, California, were randomly surveyed in late 1977 to determine which arthropods caused them the greatest problems (Frankie et al., in press). Half of those surveyed lived in

FIGURE 2.1   Nine regional divisions established by the Center for Disease Control, U.S. Public Health Service.

a lower-income neighborhood, the other half in an upper-income neighborhood. The lists obtained from these interviews were then compared with the lists obtained from the university extension services in northern and southern California. Other lists, developed from interviews with 25 pest control operators in the San Francisco Bay area (Frankie and Magowan, in press), were also used (see Table 2.4 for indoor pests and Table 2.5 for outdoor pests).

The five different groups all reported similar indoor pest problems. Ants, cockroaches, fleas, flies, and termites were repeatedly mentioned among the five chief pests, although not always in the same order. The similarity is perhaps to be expected, since the indoor environment is less variable than the outdoor environment. Among the outdoor pests, ants, aphids, earwigs, sowbugs/pillbugs, and wasps were frequently mentioned. Snails/slugs (Mollusca) were also commonly mentioned as being pests, particularly in northern California. In summary, pest lists supplied by the California extension service reflected reasonably good agreement with other representations of pest problems in the state, and particularly good agreement with regard to indoor pests.

EVALUATION OF PRINCIPAL PESTS IN DALLAS, TEXAS

Frankie and Levenson (1978) evaluated the attitudes and practices of urban dwellers in Dallas toward insect problems and insecticide use. One part of the study involved finding out which pests were the principal problems, indoors and outdoors, for lower-middle- and upper-middle-income groups. To complement this effort, a written survey was conducted of local public health officials, research and extension service personnel from Texas A&M University, and 25 pest control operators in the Dallas area. Each individual contacted was asked to rank the 10 most important indoor pests. Table 2.6 shows the responses of the different groups.

It is interesting to note that although there was general agreement that one or another pest was a problem, there was less unanimity as to seriousness. Flies, for example, were ranked by public health officials as number 10, by university research personnel as number 2, and by lower-middle-income residents as number 3. Termites were the second most prevalent problem cited by pest control operators and the third most prevalent cited by residents and researchers, but were not mentioned at all by public health officials.

Cockroaches, however, were almost unanimously deemed to be the number 1 pest problem in Dallas. The 10 most important pests (giving equal weight to each group of respondents) were: cockroaches (with an average ranking of 1.2), termites (3.8), fleas (4.5), flies (4.7), ants (5.3), spiders (7.5), ticks (7.8), crickets (8.7), mealybugs (8.8), and moths and waterbugs, which tied (9.2).

EVALUATION OF PESTS BY PEST CONTROL OPERATORS

In 1979 the National Pest Control Association asked its members to rate the importance of various pests. Responses were received from 216 members (about 10 percent of the membership) representing all geographic areas. Responses were tabulated by geographic area, though not all questionnaires could be used because of incomplete answers. To facilitate graphic presentation, response frequencies were weighted according to importance (as perceived by the pest control operator on a priority scale) and then standardized for number of respondents. This resulted in the Pest Importance (PI) index used in Figures 2.2 and 2.3. The index is only for relative comparison within each figure. [PI $= \Sigma$(no. of respondents $\times$ importance value)/no. of respondents].

A surprising element was the lack of geographic variation (see Figure 2.2). General pests (arthropods) were deemed of greatest importance by all the operators. Wood-destroying insects were next, the index for the

TABLE 2.2 Distribution of Major Indoor Pest Groups by Center for Disease Control Regions

Group	New England	Middle Atlantic	East North Central	West North Central	South Atlantic	East South Central	West South Central	Mountain	Pacific
Structural pests									
Cockroaches	+	+	+	+	+	+	+	+	+
Powder-post beetles and related beetles	+	+			+	+			+
Termites	+	+		+	+	+	+	+	+
Stored-product pests	+	+	+	+	+	+	+	+	+
Fabric and paper pests									
Clothes moths	+	+			+	+	+	+	+
Dermestidae	+	+	+	+	+	+	+	+	+
Silverfish	+		+	+		+		+	+
Nuisances									
Ants	+	+	+	+	+		+	+	+
Box elder bugs				+					
Crickets				+					
Earwigs		+				+			
"Firewood insects"									+

Fleas	+		+	+	+	+	+	+		+
Millipedes	+		+							+
Moths	+		+	+		+	+	+		+
Nonbiting flies				+		+	+			
Sowbugs/pillbugs				+		+		+		+
Spiders			+	+	+	+	+		+	+
Ticks										
Wasps		+			+	+		+		+
Public health pests										
Biting flies				+						
Fire ants				+						
Fleas	+		+		+	+	+	+	+	+
House dust mites										
Spiders			+	+	+	+	+	+	+	+
Ticks					+	+	+	+		
Wasps		+								+
Houseplant pests										
Mites	+		+		+	+	+	+	+	+
Mealybugs		+			+	+	+	+	+	+
Scales					+	+		+		+
Whiteflies		+								+

NOTE. Based on NRC Committee on Urban Pest Management study of extension services in all land grant institutions. (See Appendix D.)

TABLE 2.3 Distribution of Major Outdoor Pest Groups by Center for Disease Control Regions

Group	New England	Middle Atlantic	East North Central	West North Central	South Atlantic	East South Central	West South Central	Mountain	Pacific
Plant-eating pests									
Chewing									
Leaves & stems	+	+	+	+	+	+	+	+	+
Roots	+	+	+		+	+			+
Boring-tunneling		+	+			+			+
Piercing-sucking	+	+	+	+	+			+	+
Miscellaneous		+	+						+
Nuisances									
Ants	+				+		+	+	+
Box elder bugs		+		+					
Crickets									
Earwigs							+	+	+
Fleas						+			+

Millipedes

Moths

Nonbiting flies

Snails/slugs

Sowbugs/pillbugs

Ticks

Wasps and bees

Public health pests

  Biting flies

  Chiggers

  Fire ants

  Fleas

  Ticks

  Wasps and bees

Structural pests

  Termites

NOTE. Based on NRC Committee on Urban Pest Management study of extension services in all land grant institutions. (See Appendix D.)

TABLE 2.4 Major Indoor Urban Arthropod Pests in California, as Perceived by Homeowners, Pest Control Operators (PCOs), and Extension Service Personnel

Order of Importance	Northern California				Southern California
	Lower-income Homeowners[a]	Upper-income Homeowners[a]	Extension Entomologists	PCOs[b]	Extension Entomologists
1	Ants	Ants	SPP[c]	Cockroaches	Ants
2	Fleas	Fleas	Cockroaches	Fleas	Flies
3	Flies	Flies	Fleas	Silverfish	SPP[c]
4	Cockroaches	Termites	Termites	Ants	Fleas
5	Moths	Moths	Ants	Termites	Cockroaches
6	Spiders	Cockroaches	Powder-post beetles[d]	SPP[c]	Earwigs
7	Mites	Spiders	Spiders	Spiders	Spiders
8	Mealybugs	Mealybugs	Flies	Fabric pests	Termites
9	Earwigs	Silverfish	"Firewood insects"[e]	Sowbugs	Crickets
10	Termites	Mites and wasps	Fabric pests	Bees, wasps, and wood borers	Moths

NOTE. Brackets indicate ranks of equal frequency.

[a] From each income group, 100 homeowners in the city of Berkeley were randomly interviewed. (Levenson and Frankie [In press]).

[b] Twenty-five PCOs were interviewed in the San Francisco Bay Area (Frankie and McGowan [In press]).

[c] Stored-product pests.

[d] Pests not prioritized in this column past rank 5.

[e] Insects associated with woodpiles.

TABLE 2.5 Major Outdoor Urban Arthropod Pests in California, as Perceived by Homeowners, Pest Control Operators (PCOs), and Extension Service Personnel

Order of Importance	Northern California				Southern California
	Lower-income Homeowners[a]	Upper-income Homeowners[a]	Extension Entomologists	PCOs[b]	Extension Entomologists
1	Snail/slugs	Snail/slugs	Aphids	Ants	Flies
2	Ants	Aphids	Whiteflies	Sowbugs	Ants
3	Aphids	Wasps and bees	Bark beetles	Earwigs	Aphids
4	Mosquitoes/flies	Ants	Ants	Aphids	Cutworm larvae
5	Fleas	Moths	Wasps	Spiders	Earwigs
6	Termites	Leafhoppers	Snails[c]	Oakworm	Wasps
7	Wasps	Mosquitoes/flies	Earwigs	Mites	Whiteflies
8	Pillbugs	Worms	Sowbugs	Scales	Mosquitoes
9	General pests	Termites	Oakworm	Turf pests	Crickets
10	Earwigs	Sowbugs/earwigs	Root maggots	Crickets	Grasshoppers

NOTE. Brackets indicate ranks of equal frequency.

[a] From each income group, 100 homeowners in the city of Berkeley were randomly interviewed. (Levenson and Frankie [In press]).

[b] Twenty-five PCOs were interviewed in the San Francisco Bay Area (Frankie and McGowan [In press]).

[c] Pests not prioritized in this column past rank 5.

TABLE 2.6 Rank in Importance of Indoor Urban Pests in Dallas, According to Residents, Public Health Officials, Texas A&M University Research and Extension Personnel, and Pest Control Operators (PCOs)

Pest	Dallas Residents[1]		Public Health Officials[2]	Texas A&M University[2]		25 PCOs[3]	Average Rank n=6
	Lower-middle-income	Upper-middle-income		Research	Extension		
Ants—general	2	2			3	3	5.3
Ants—carpenter				4			9.8
Aphids				9			10.7
Bedbugs			9				10.7
Carpet beetles (dermestids)					5		10.0
Centipedes			8				10.5
Chiggers & other biting mites			4				9.8
Cockroaches—general	1	1	1	1	2	1	1.2
Crickets	7c	6d		8		9	8.7
Earwigs		5b					10.0
Fleas	5a	4a	3	5	6	4	4.5
Flies	3b	5a	10	2	8		4.7
Lice			5				10.0
Mealybugs	7a	6b		7			8.8

Pest							Mean
Moths (misc. indoor flying)	5b	6a					9.2
Pillbugs/sowbugs						10	10.8
Rodents (rats & mice)		6a			10	6	10.2
Powder-post beetles					9	7	10.7
Silverfish	6			10			9.3
Scorpions			7				10.3
Spiders	4	5c	6			8	7.5
Spider mites				6			10.2
Stored-food insects					4	10	9.7
Termites–general	3a	3		3	1	2	3.8
Ticks			2		7	5	7.8
Waterbugs	7b	4b					9.2
Weevils		6c					10.2
Other pests	8	7					9.8

NOTE. On a scale of 1 to 10, 1 is most important, 10 is least; "a" is most important, "d" is least. Although waterbugs are cockroaches, they are indicated here as reported by respondents.

SOURCE:
[1] G. W. Frankie, University of California, personal communication, 1979.
[2] T. A. Granovsky, Texas A&M University, personal communication, 1979.
[3] Granovsky and Frankie (In press).

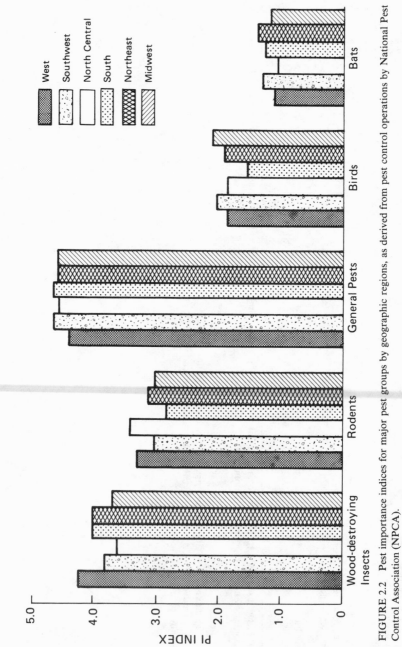

FIGURE 2.2 Pest importance indices for major pest groups by geographic regions, as derived from pest control operations by National Pest Control Association (NPCA).

SOURCE: National Pest Control Association (Unpublished).

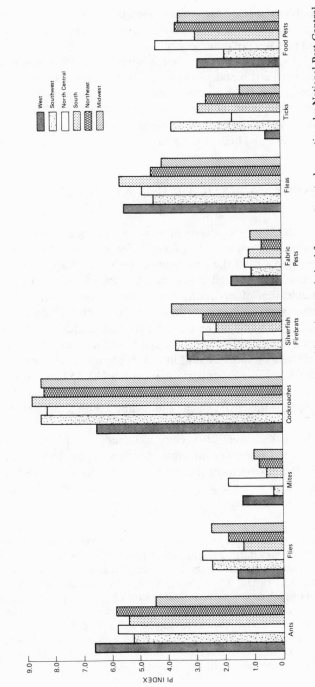

FIGURE 2.3 Pest importance indices for arthropod pests by geographic regions, as derived from pest control operations by National Pest Control Association (NPCA).

SOURCE: National Pest Control Association (Unpublished).

western states being slightly higher than for other areas. Rodents were a close third, and house mice were deemed more important than Norway rats. Birds and bats were of minor concern.

As Figure 2.3 reveals, however, some geographic diversity was evident in the judgments of arthropod pests alone. Cockroaches, followed by ants and fleas, were held to be of greatest concern, although cockroaches were less important in the West, where ants and fleas (but not ticks) are more prevalent. Ticks and silverfish stood out as problems in the Southwest. Food pests and flies rated high in the Midwest.

URBAN PESTS AND PESTICIDE USE: ATTITUDES AND BEHAVIOR

There has been much recent interest in social-psychological variables with reference to the environment in general. For example, the Scientific Committee on Problems of the Environment (SCOPE), a special committee of the International Council of Scientific Unions, has initiated a project on the communication of environmental information and societal assessment and response. The Neighborhood Environmental Evaluation and Decision System (NEEDS) in the U.S. Department of Health, Education, and Welfare was designed to help the Department recognize the effect of environmental stress on individual health, both physical and mental (CEQ 1971). And the American Psychiatric Association (APA) has appointed a Task Force on Ecopsychiatric Data Base that has reported on the relationship between the environment and mental health and illness (APA 1979).

However, there has been little empirical study of the psychosocial impacts of pests and pesticides on urban dwellers. Some theories suggest that stimuli in the environment like pests or pesticides have an effect on mental health, social well-being, and the general satisfaction of inner city residents. Rene Dubos wrote:

The widespread belief that man can adapt or get used to anything constitutes one of the main difficulties in evaluating the impact of pollution on mental health and in studying the mechanisms of its effects. . . . But this does not mean that such tolerance is desirable. In fact it is often achieved at the cost of some loss in the desired attributes which do constitute mental health. The most deplorable aspect of existence in American cities may not be murder, rape and robbery, but the constant exposure of children to pollutants, noise, ugliness and garbage in the streets. This constant exposure conditions children to accept public squalor as the personal state of affairs and therefore handicaps them mentally at the beginning of their lives (National Institute of Mental Health 1972).

The APA (1979) has recently issued a bibliography dealing with the effects of natural and man-made environments on mental health. Inspection of these volumes reveals little specific study of the psychosocial impacts of pests or pesticides: Of 604 references in the APA bibliography, only 1 reference deals with a subject in this area (i.e., Gershon and Shaw 1961, "Psychiatric sequelae of chronic exposure to organophosphate insecticides"). Similarly, a bibliography prepared by the NIMH (1972) contains only 2 references concerning urban pesticide use.

Data on the impact of pesticides on psychosocial variables are also rare and often not well defined. Gershon and Shaw (1961) found that people exposed to organophosphate insecticides for 1 to 10 years showed schizophrenic and depressive reactions with severe memory impairments. However, in an epidemiological survey Stoller et al. (1965) found no relationship between the sale and use of organophosphate insecticides and the incidence of major mental illness.

In conclusion, psychosocial impacts from pests and pesticides are suggested by theory, observation, and several empirical studies, but there are few clear-cut data to support the associations. Part of the reason stems from the nature of the variables being studied. They are difficult to document, not because they are irrelevant or insignificant, but because they are multifaceted, confounded with other variables, and embedded within the overall subject of residential housing and neighborhood quality.

What are urban dwellers' perceptions of what constitutes a pest? How do people decide when to control a pest? Are some "pest problems" actually people problems? Are certain variables such as sex, socioeconomic status, or culture, differentially associated with pesticide use? Does specific environmental knowledge make a difference in how people control pests? Are there social or psychological barriers to the widespread dissemination of new pest control strategies, such as integrated pest management? The following sections examine what we now know about the attitudes and behavior of people toward pests in urban environments, and what we need to know.

LITERATURE REVIEW

Environmental Attitudes

Over the past decade, interest in people's attitudes toward the environment in general and "environmental quality" in particular has grown substantially. Dunlap and Van Liere (1978) have appraised the existing literature and make this generalization:

The best predictors of environmental concern are age, education, and political ideology, with the young, the well-educated, and the politically liberal being more environmentally concerned than their counterparts. . . . Other variables which have been found to correlate with environmental concern, though with less consistency and/or with lower magnitude include residence, race, income, and political party preference (with urban residents, Caucasians, higher income groupings and Democrats ranking higher in environmental concern). . . . While our knowledge of the effects of demographic variables on environmental concern is limited, our knowledge of the effects of social-psychological and personality variables . . . appears even more limited. To some extent this reflects the fact that there have been fewer studies examining the latter types of variables . . . than there have been of studies examining demographic variables. Therefore, even though a number of studies . . . have investigated the relationship between environmental concern and social-psychological variables, we are unwilling to regard the results as established empirical generalizations until they are confirmed by additional research. To some extent, our reluctance also stems from the fact that many of these studies were carried out with very restrictive samples, not only for limiting the generalizability, but also making it impossible for researchers to take into account these facts of demographic variables known to be related to environmental concern.

Attitudes and Behavior Toward Pests and Pesticides

Salcedo et al. (1971) cited a report based on 3,800 interviews showing that 92 percent of those sampled thought pollution was a problem. Pesticides were ranked high on the list of pollutants that people thought should be banned. But despite considerable publicity and widespread use of pesticides, investigation of attitudes and behavior toward pests and pesticides has been minimal.

Dunlap and Van Liere (1978) list only two studies dealing with public attitudes toward pests and pesticides. While other studies do exist, the number is quite small, and most of these pertain to the agricultural environment. The Working Group on Environmental Perception of the International Geographical Union has a newly formed research program on "Perception and Management of Pests and Pesticides," but the program is geared toward agricultural pests, especially in the developing countries.

There are two notable studies among the few that compare rural and urban residents' attitudes toward pesticides. Salcedo et al. (1971) surveyed 101 farmers and 199 city dwellers in Champaign County, Illinois. Respondents, who differed in educational level and in the amount of money they spent on pesticides, were presented with 10 statements about the pesticide industry. A great majority of both groups (82 percent) had used pesticides during the 6 months before the interview. Some 43 percent of all respondents held favorable attitudes toward the pesticide industry, while 38 percent held unfavorable attitudes. The responses also revealed that more city dwellers than farmers were concerned about the harmful effects of pesticides. City dwellers also expected the pesticide industry to make a greater effort to educate the public on pesticide safety.

Ryan et al. (1974) also examined rural and urban differences in attitudes and knowledge about the use of insecticides. The target populations were homeowners in Tucson and commercial farmers in two Arizona counties. The survey showed that farmers felt there was more of a need for insecticides than did the city dwellers. Although the farmers were more aware of the hazards of insecticides, the urban group was more concerned about the potentially adverse effects of pesticide residues on health. This is consistent with the findings of the Salcedo et al. study. Farmers had more specialized sources of information, and their responses reflected a high degree of knowledge about insecticide properties. In general, the higher the level of education, the more knowledgeable people were about insecticides. The results suggest that agricultural extension services have had little impact on education about pesticides in the urban sector. The authors emphasize that differences between the responses of farmers and urban dwellers underscore the need for different approaches in communicating with the two groups.

The following empirical studies are particularly informative.

Olkowski et al. (1979) point out that urban pest problems and control practices may be distinctive because (a) some pests increase in number as a direct result of the crowded living conditions and deteriorated buildings typical of inner-city areas; (b) entirely new pest problems sometimes ensue when urban areas expand into formerly rural areas; and (c) cosmetic standards may be higher in urban environments, resulting in more extensive pest control efforts.

The literature on urban dwellers' attitudes and behavior toward pests and pest management is considered below.

Lande (1975), in a study of public attitudes toward pesticides in Allegheny County, Pennsylvania, found widespread household use (85 percent) among the residents of the 39 single-family dwellings covered by the survey. Lande notes, however, that "many householders were not aware of all the pertinent information on pesticide labels." In deciding which pesticides to buy, most users did not obtain expert aid or information. Instead, they relied on advertisements or product availability. These findings corroborate those of von Rumker et al. (1972), who found that the majority of pesticides are purchased on the basis of information provided by the manufacturer or the retailer. Lande (1975) concluded that because pesticides were reported to constitute less than 5 percent of all poisonings, pesticides are neither "unusually dangerous among all household chemicals nor among hazardous chemicals present in urban environments."

Finklea et al. (1969) investigated the use of pesticides in a sample of urban South Carolina households because "acute pesticide poisoning has . . . caused substantial morbidity and mortality among South Carolina children, particularly Negro children." White (121) and black (75) families were sampled. The investigation showed that 89 percent of all the families used household pesticides, a finding consistent with Lande's results. Black families used pesticides more frequently than did white families (97 percent versus 83 percent). The authors hypothesized that the difference existed because the homes of the black families were more infested with flies, cockroaches, and mosquitoes. The investigators concluded that racial differences in pesticide use were "probably not real, but rather reflected economic differences in housing quality and vocation."

Other findings of the study were that private pest control operators regularly treated the residences of 42 percent of the families covered by the survey, and that household pesticide purchases were most frequently made at grocery stores. Common-sense safety precautions were largely ignored by both white and black families, with 88 percent failing to lock pesticides away, 66 percent storing them within easy reach of small children, and 54 percent placing them near food or medicine. The researchers pointed out that, fortunately, industry and regulatory agencies have done much to protect consumers from their nonchalant use of household pesticide preparations. However, Finklea et al. (1969) point out that many yard and garden pesticide preparations that require further dilution by the consumer may be quite toxic.

The Minnesota-Wisconsin EF-USDA Agricultural Home Horticultural Project (Keel et al., undated) was designed to survey home gardeners in a metropolitan area. The study was prompted by the increasing demand for agricultural extension services among amateur gardeners in city environments. The research effort indicated that nearly all the households (94 percent) in the Minneapolis-St. Paul metropolitan area raised some type of plant. Friends, neighbors, and family were frequent sources of information about pests and pest control for more than two-thirds of the respondents. Stores selling home garden products were used by 20 percent of the gardeners as sources of information. It was found that those who usually contacted extension services for information were younger, better educated, and lived in higher-income households. Although the study is a reasonably good attempt to determine sources of information on gardening practices, it does not indicate what types of information were transmitted, nor is any mention made of health problems that might be related to garden pest management.

A national household pesticide use study was conducted in 1976-1977 by the Epidemiologic Pesticide Study Center of Colorado State University in cooperation with nine other epidemiologic pesticide study centers and projects (Savage 1978). Although its results were reported in terms of 10 regions in the United States and therefore cover both urban and rural households, the data are summarized here because this is one of the few studies to gather information regarding the quantities of pesticides used by householders. The overall data on pesticide use from 8,254 interviewees are consistent with those found in other studies. Nine out of every 10 residents interviewed reported using some type of pesticide in their house, garden, or yard:

Over three times as many householders use pesticides in their houses than in their yards. Such widespread use of pesticides in the home environment is undoubtedly a significant source of exposure of the general population to pesticides.

People living in the southeastern United States were found to have greater possibilities of exposure to termiticides and other pesticides used by professional pest control operators than people living in other parts of the country. Many of the respondents were not aware of what pesticide they used. Less than 6 percent went to knowledgeable sources for pest control information. Despite the fact that only 3 percent (253) of the respondents were reported to have suffered illness (e.g., nausea, dizziness, headaches) as a result of pesticide use, the study concludes that "household use of pesticides may have a more significant role in human exposure to pesticides than previously thought."

The studies discussed above focused chiefly on behavior or on behavior and one component of attitudes (e.g., information). They did not attempt to relate actual behavior to attitudinal data or to relate important variables (e.g., gender, socioeconomic status, profession) with attitudes or behavior. A recent study by Levenson and Frankie (in press) on homeowner practices and attitudes toward pests and pesticides in three metropolitan areas has sought to fill the gaps. The study is also one of the few to examine a previously ignored component of attitudes, namely, affect, or feeling. How do people feel about insects? Are there some insects that people like? How are a person's affective responses related to pesticide use? The Frankie and Levenson study contains a description of attitudes and behavior over a 3-year time period in one location in order to assess changes over time, and tries to establish when and why people use insecticides.

In 1977-1978, data were collected from 601 adults in Dallas, Texas;

Berkeley, California; and New Brunswick, New Jersey. Interviewers used a structured questionnaire in door-to-door surveys. Questionnaire items were designed to permit an evaluation of the affective (emotional), behavioral, and cognitive components of attitudes. The typical person surveyed was a white, married adult who had owned a house in either an upper-income or lower-income neighborhood for approximately 17 years. He or she was about 44 years old and was married to someone in a semiprofessional occupation. People meeting certain demographic characteristics were sampled in a modified quota-sampling procedure in order to facilitate group comparisons (e.g., differences based on race, gender, socioeconomic status).

The overall response indicates that insect problems are quite widespread. Of the 601 respondents, 539 said that they had an indoor or outdoor pest problem. The most frequent source of information about pest problems was pest control operators, and people felt generally satisfied with this information. Although most respondents said they had changed their attitudes and were now more cautious in using pesticides, they rarely knew the type of chemical used by the professionals they employed. Nonchemical means of controlling pests were deemed less satisfactory than pesticides. The responses to questions on when and why they used insecticides revealed significant regional and socioeconomic differences. For example, lower-income residents in Berkeley used chemicals to prevent the occurrence of outdoor pests, while upper-income respondents waited until they observed unwanted pests before making use of pesticides. The people interviewed in Dallas, however, responded in an opposite manner, with wealthier residents using more preventive measures. Although the respondents felt that their attitudes toward pesticide use were changing, pesticides were given high marks as a means of pest control. Over half of the respondents also said they liked some insects, chiefly because of their utilitarian value.

The survey showed few differences in response based on gender, marital status, race, or even type of neighborhood. Although it was expected that lower-income neighborhoods would have more pest problems than upper-income ones, respondents from both kinds of areas reported the same incidence of indoor and outdoor pest problems. Homeownership, however, was an important variable. Homeowners reported more outdoor pest problems than renters, while the latter group reported more indoor pest problems.

A consistent picture was revealed by the analyses: People who reported having pest problems were more apt to like insects and to be more aware of beneficial insects. The authors of the study hypothesize that a general sensitivity to insects may result in people being more aware of both helpful insects and those that cause problems. This finding may have implications for educational efforts in pest management. Such educational efforts may have the unintended effect of heightening people's sensitivity to perceive pests.

The Levenson and Frankie study (in press) also evaluated changes in the incidence of pest problems and related attitudes and behavior over time. Information obtained from residents of Dallas in 1978 was compared with responses from different Dallas residents in 1975 and 1976. In general, it appears that Dallas residents had become less satisfied with the information they were obtaining on pest problems. The 1975 sample reported more outdoor pest problems, more use of professional exterminators, more chemical use, and more knowledge of beneficial insects than did those sampled later.

Another particularly informative study regarding household use of pesticides and related knowledge, attitudes, and behavior is that conducted by Drummond and Mood (1973). The study involved people who were the target population of an urban rat control project in Philadelphia. Of those in the target population, 82 percent belonged to no local community organization, 35 percent had never been to high school, and 9 percent were Spanish-speaking. The study was unusual in four respects: (1) an inspection of the premises to ascertain environmental problems was made at the time of the survey; (2) the study was designed to assess the effectiveness of a project to educate citizens; (3) actual participation in neighborhood cleanup action was noted; and (4) the study focused on inner-city residents. The results of the study, based on 1,224 interviews, suggest that people will participate in a project that they perceive to be immediately beneficial, although they will not necessarily go to regular meetings on the subject. Interpersonal communication seemed to be the most effective way to promote the project. Proper refuse disposal followed economic lines. Of people with incomes under $4,000, only 26 percent used metal garbage cans, compared with 42 percent of those with incomes over $7,000. (The use of metal cans, however, was unrelated to objective evidence of rats on the premises.)

Citizens involved in the project were relatively successful in persuading landlords to make repairs related to pest management, although Spanish-speaking people had poorer success than blacks or whites. Interestingly, the data suggest that those who call a city government for help have greater environmental problems than those who do not. "This result tends to contradict the idea that many complaints come from vocal persons and groups with few problems while greater problems pass unnoticed, a concept occasionally promoted by agencies which do not wish to provide complaint-oriented services" (Drummond and Mood 1973).

The studies discussed above suggest the following generalizations: (a) household pesticide use is quite common; (b) people are ill-informed about the harmful effects of pesticides and rarely take the recommended precautions; (c) pest control operators are widely used as a source of information about pesticide application; and (d) people believe pesticides do good and rarely think they do harm. It also appears that despite widespread pesticide use, many people actually like certain types of insects. The main social factors that seem to influence pest management attitudes and behavior are income and homeownership.

Only a few studies have investigated the attitudes and behavior of professional pest control operators or retail sellers of pesticides. Their feelings, knowledge, and behavior are important because their influence on the extent of pesticide use is sizeable. As indicated previously, urban dwellers rely heavily on such people, either for information or for the application of pesticides.

As an example of the attitudes and behavior of retail sellers of pesticides, Beal et al. (1969) completed a series of studies in Iowa, one of which pertains to urban pesticide dealers' attitudes, knowledge, and perception of their roles. The researchers focused on several different kinds of chemical pesticides: those used on lawns and gardens, those used to kill rodents, those used to eliminate other household pests, and those used to kill the pests that harm ornamental indoor plants. Interviews were conducted with 150 urban pesticide sellers, many of whom worked in grocery stores (25 percent) or hardware stores (16 percent). The others were employed in drugstores, grocery stores, variety and other stores.

Household pesticides comprised the largest percentage of sales. Personal data showed that more than half of the sellers had not graduated from high school. When asked to answer questions about pesticides, the sellers were usually accurate about proper safety precautions (90 percent) and knowledge of regulations and controls (80 percent); however, they were considerably less informed about frequently used pesticides or commonly used chemical pesticide terminology (57 percent).

Twenty-three percent of the sellers felt pesticides were dangerous if not used properly. Almost 75 percent agreed with the statement that "anyone with a home or garden should use chemical pesticides to control insects and disease."

A unique feature of this study was its exploration of the sellers' self-perceived roles. Almost half felt that customers expected them to be a source of information on pesticides, and an equal percentage thought they should make recommendations to the potential user.

ATTITUDES AND BEHAVIOR INVOLVING INTEGRATED PEST
MANAGEMENT

Although a recent report by the Council on Environmental Quality (CEQ
1978) states that "a promising alternative to the use of pesticides for the
control of pests is the concept of integrated pest management," IPM does
not exclude the use of pesticides; it is designed to use all available
techniques after consideration of the pests' damage potential and the
ecological, social, psychological, and economic costs involved. The
approach considers the entire ecosystem (Huffaker et al. 1978). President
Carter's 1979 environmental message directed appropriate federal agencies
to modify existing research, control, and assistance programs "to support
and adopt IPM strategies wherever practicable" (The White House 1979).

To what extent will attitudes and behavior make a change to IPM
feasible and efficient or impractical and difficult? Andrews (1974) has this
to say:

In order to make more widespread use of [IPM] knowledge we must change the
direction of conventional thinking on pest control. The IPM concept incorporates
the philosophical underpinnings of such a change in attitude. . . . Implementation of this approach has as much to do with reorienting public attitudes and
everyday behavior as it does with technique, knowledge, and the actions of
governmental agencies.

Although there is a massive amount of psychological literature on such
things as the content of a message, the characteristics of its audience, and
the importance of the message's source in affecting attitude change
(McGuire 1969), we know very little about the public's possible adoption
of IPM. What is available are descriptions of the psychological obstacles to
acceptance of IPM techniques by institutional bureaucracies and reports
on the results of other social change efforts. These are briefly examined
here to emphasize the importance of social-psychological factors in the
implementation of new programs.

Olkowski et al. (1978) have described three main psychological barriers
to the spread of IPM technology. The first is that of inertia: "Adopting an
IPM approach . . . usually means instituting some change. . . . It is
always easier to continue what one has always done." The second may
come from the implication that if a change is necessary, past performance
was poor, and, conversely, that "By protesting IPM as nothing new, the
speaker is defending past performance." The third is anticipated personal
losses. The introduction of new techniques may lead people to fear that
they may lose their jobs and, subsequently, their personal and economic
status.

The authors recommend additional methods to reduce resistance to the adoption of IPM. They suggest disseminating information about IPM to such organizations as community garden groups and consumer groups, and setting up forums to exchange experiences. "The sometimes bitter experience of the past years has taught the lesson that improvement in the daily environment of people must be sought with their help and guidance, not imposed on them by those who claim to know what is best" (CEQ 1971).

For the urban dweller who has immediate needs and expectations regarding pest control, improvement of the "balance of nature" or even of his or her own health may not be viewed as more important than eradication of pests, using the fastest, cheapest way possible. Urban dwellers with certain kinds of cultural and educational backgrounds may find it difficult to delay the immediate gratification offered by pesticides in controlling pests in order to achieve a greater future benefit. Such delay requires an ability to understand that the rewards of delay are often not obtained personally but shared by the community, that the risks of pesticides are often not personal risks but hazards to the natural environment, and that the harmful effects of pesticides are often not immediately evident. Frankie and Levenson (1978) found that people who had personally observed a negative effect from pesticides (e.g., a child's illness, death of a pet) were more cautious in using pesticides than people who had not personally witnessed such results.

RESEARCH PRIORITIES

There appear to be a number of major deficiencies in the psychological research pertaining to pests and pest management. While everyone seems to agree about the importance of such research in understanding urban dwellers' attitudes and behavior, little work has actually been undertaken. Part of the reason for the sparsity of data may be the practical, theoretical, and personal difficulties of employing truly multidisciplinary approaches to problems. Following are eight research priorities for improving the situation.

1. The first priority is to increase the amount of basic information on the frequency, quality, and type of pesticide use within the urban environment. Baseline data are also needed on perception of "pests" by inner-city residents and types of pest problems experienced by city dwellers. Information is needed on the social, situational, personal, and attitudinal factors involved in pest perception and pest management. Also

needed are data on race, gender, homeownership, and other factors, which quite often can be easily obtained.

2. The second priority is to develop useful theories to guide research more efficiently. Theories already exist on attitude change processes, perceptual differences based on personality factors, and the situational effects that influence behavior. Urban researchers and professionals concerned with pest management could use existing paradigms to develop their own models in order to advance empirical knowledge. For example, Tait's study (1979) of the way farmers perceive pests and disease is designed to test a formula to predict behavior from attitudinal data.

3. Appropriate methodology must be further developed, and information acquired by means other than surveys. Inner-city dwellers, especially those from different cultures or with minimal educational backgrounds, are often not able to respond to questions put to them by interviewers of a different race or socioeconomic background. Some of the alternative techniques mentioned by Whyte (1977) for use in developing countries (such as observation and projective techniques) might be useful in this regard. In addition, more reliable techniques should be devised to assess such qualitative concepts as quality of life and pest-perception threshold. When attitudes are assessed, their different attributes need to be ascertained (e.g., strength, direction, type, consistency). Furthermore, related constructs, such as beliefs and values, require study.

In addition, more studies are needed that analyze the relationships among important variables, and the different levels of a single variable. The one-question study and the univariate analysis of data should be replaced by multidimensional assessment techniques and multivariate statistical designs (e.g., factor analysis, multiple regression equations). These kinds of analyses more accurately reflect the complex relationships of the "real world" and will lead to further improvements in the accuracy and usefulness of data.

4. Accurate conceptualization of what constitutes a desired outcome will permit the development of more realistic programs devoted to changing attitudes and behavior. Such programs should be developed to facilitate the correlation of conditions found in previous studies with preferred behavior. Assessment should also permit other possible outcomes in addition to the desired one. Would stricter pesticide regulation, for example, have undesirable effects on minority groups?

5. Previous studies on pests and pesticide use have concentrated on the independent consumer (farmer, urban dweller). Additional work on the attitudes and behavior of pesticide sellers, pest control operators, community organizations, government officials, and extension agents will be crucial in determining the future effectiveness of urban pest management.

6. There is a need for different approaches to research and training. Research on pest management must involve multidisciplinary teams consisting of social scientists, economists, entomologists and other pest management professionals, and ecologists with appropriately broad interests and skills. In addition, others such as political scientists, psychologists, anthropologists, philosophers, and agronomists could also make valuable contributions. In order to encourage the formation of such research teams, existing research arrangements will have to be revised.

7. Alternative graduate training programs should be designed and implemented to prepare scientifically trained people for multidisciplinary, holistic analyses. The training of urban entomologists, for example, could include courses and experience in urban planning, attitude measurement, and behavioral analysis.

8. An additional research priority is to conduct more homeowner surveys in which inventories of public-health and nonpublic health pests are made. It is important that comparative studies between lower- and upper-income neighborhoods be made in representative U.S. cities.

REFERENCES

Alcamo, I.E. and A.M. Frishman (1980) The microbial flora of field-collected cockroaches and other arthropods. Journal of Environmental Health 42(5):263-266.

American Psychiatric Association (1979) American Psychiatric Task Force Report for Mental Illness. Washington, D.C.: American Psychiatric Association.

Anderson, A.E., P.B. Cassady, and G.R. Healy (1974) Babesiosis in man. Sixth documented case. American Journal of Clinical Pathology 62(5):612-618.

Andrews, F. (1974) Social indicators of perceived life quality. Social Indicators Research 1:279-299.

Anonymous (1976) Lice-yes, lice again a problem. Journal of Infectious Diseases 6:21.

Bahmanyar, M. and W. Cavanaugh (1976) Plague Manual. Geneva: World Health Organization.

Barnard, J.H. (1973) Study of 400 Hymenoptera sting deaths in the United States. Journal of Allergy and Clinical Immunology 52:259-264.

Beal, G.M., J.M. Bohlen, G.W. Edwards, and W.A. Fleischman (1969) Behavior Studies Related to Pesticides. Ames, Iowa. Cooperative Extension Service. Iowa State University.

Beneson, A.S., ed. (1975) Control of Communicable Diseases in Man. 12th ed. New York: American Public Health Association.

Bernton, H.S. and H. Brown (1964) Insect allergy: Preliminary studies of the cockroach. Journal of Allergy 35:506-513.

Bernton, H.S. and H. Brown (1970a) Cockroach allergy: Age of onset of skin reactivity. Annals of Allergy 28:420-422.

Bernton, H.S. and H. Brown (1970b) Insect allergy: The allergenicity of the excrement of the cockroach *Blattella germanica*. Annals of Allergy 28:543-547.

Blommers, L. and M. van Lennep (1978) Head lice in the Netherlands: Susceptibility for insecticides in field samples. Entomological Experiments and Applications 23:243-251.

Bowen, G.S., T.R. Fashinell, P.B. Dean, and M.B. Gregg (1976) Clinical aspects of human Venezuelan equine encephalitis in Texas, 1971. Bulletin of the Pan American Health Organization 10:46-57.

Brooks, J.E. (1973) A review of commensal rodents and their control. Critical Review of Environmental Control 3(4):405-453.

Cornwell, P.B. (1968) The Cockroach. Vol. 1. London: Hutchinson.

Council on Environmental Quality (1971) Environmental Quality. Second Annual Report. Washington, D.C.: U.S. Government Printing Office.

Council on Environmental Quality (1978) Environmental Quality. Ninth Annual Report. Washington, D.C.: U.S. Government Printing Office.

Drummond, D.W. and E.W. Mood (1973) Actions of residents in response to environmental hazards in the inner city. American Journal of Public Health 63:335-340.

Dunlap, R.E. and K.D. Van Liere (1978) Environmental Concern: A Bibliography of Empirical Studies and Brief Appraisal of the Literature. Public Administration Series No. P44. Monticello, Ill.: Vance Bibliographies.

Elzweig, M.A. and A.M. Frishman (1977) The role of the PCO in controlling human lice. Pest Control 45(11):33-34.

Feigin, R.D., L.A. Lobes, Jr., D. Anderson, and L. Pickering (1973) Human leptospirosis from immunized dogs. Annals of Internal Medicine 79(6):777-785.

Finklea, J.F., J.E. Keil, S.H. Sandifer, and R.H. Gadsden (1969) Pesticides and pesticide hazards in urban households. Journal of the South Carolina Medical Association 65:31-33.

Frankie, G.W. and H. Levenson (1978) Insect problems and insecticide use: Public opinion, information, and behavior. Pages 359-399, Perspectives in Urban Entomology, edited by G.W. Frankie and C.S. Koehler. New York: Academic Press, Inc.

Frankie, G.W., H. Levenson, and S. Mandel (In Press) Homeowner pesticide usage in three U.S. metropolitan areas (tentative title). Hilgardia.

Frankie, G.W. and C. Magowan (In Press) A preliminary study of attitudes and practices of pest control operators toward pests and pesticides in selected urban areas of California and New Jersey. Hilgardia.

Fraser, D.W., J.I. Ward, L. Ajello, and B.D. Plikaytis (1979) Aspergillosis and other systemic mycoses. Journal of the American Medical Association 242(15):1631-1635.

Garcia, M.G. and J.A. Najera-Morrondo (1973) The interrelationships of malaria, agriculture, and the use of pesticides in malaria control. Boletin de la Oficina Sanitaria Panamericana 6(3):20.

Georghiou, G.P. (1972) Studies on resistance to carbamate and organophosphate insecticides in *Anopheles albimanus*. American Journal of Tropical Medicine and Hygiene 21:797-806.

Gershon, S. and F.H. Shaw (1961) Psychiatric sequelae of chronic exposure to organophosphate insecticides. Lancet 7191:1371-1374.

Granovsky, T.A. and G.W. Frankie (In Press) Texas pest control operators' attitudes and reactions toward pests and pesticides in Dallas, Texas, summer 1978. Hilgardia.

Guthrie, D.M. and A.R. Tindall (1968) The Biology of the Cockroach. New York: St. Martin's Press.

Hattwick, M.A., A.H. Peters, M.B. Gregg, and B. Hanson (1973) Surveillance of Rocky Mountain spotted fever. Journal of the American Medical Association 225(11):1338-1343.

Healy, G.R., A. Spielman, and N. Gleason (1976) Human babesiosis: Reservoir of infection on Nantucket Island. Science 192:479-480.

Huffaker, C.B., C.A. Shoemaker, and A.P. Gutierrez (1978) Current status, urgent needs, and future prospects in integrated pest management. Pages 237-262, Pest Control Strategies, edited by E.H. Smith and D. Pimentel. New York: Academic Press, Inc.

Jackson, W.B. and P.P. Meier (1955) Dispersion of marked American cockroaches from sewer manholes in Phoenix, Arizona. American Journal of Tropical Medicine and Hygiene 4(1):141-146.

Jackson, W.B. and P.P. Meier (1961) Additional studies of dispersion patterns of American cockroaches from sewer manholes in Phoenix, Arizona. Ohio Journal of Science 61(4):220-226.

Kappus, K., M. Moore, and M. Pollack (1979) Surveillance for human arbovirus infection, United States, 1979. Arthropod-borne Virus Information Exchange 36:157-159.

Keel, V.A., H.P. Zimmerman, and R.A. Wearne (undated) Communicating Home Garden Information. Phase I Report. The Minnesota-Wisconsin EF-USDA Agricultural Home Horticultural Project. Extension Service. Minneapolis, Minn.: University of Minnesota.

Lamizana, M.T. and J. Mouchet (1976) La pediculose en lilieu scholaire dans la region parasienne. Medecine et Maladies Infectieuses 6:48-52.

Lande, S.S. (1975) Public attitude toward pesticides, a random survey of pesticide use in Allegheny County, Pa., 1975. Public Health Reports 90(1):25-28.

Levenson, H. and G.W. Frankie (In Press) A study of homeowner attitudes and practices toward arthropod pests and pesticides in three U.S. metropolitan areas. Hilgardia.

Mann, J.B., W.J. Martone, and J.M. Boyce (1979) Endemic human plague in New Mexico: Risk factors associated with infection. Journal of Infectious Diseases 140(3):397-401.

McGuire, W.J. (1969) The nature of attitude and attitude change. In Handbook of Social Psychology, edited by G. Lindzey and E. Aronson. Vol. 3. Reading, Mass.: Addison-Wesley.

Marks, G. and W.A. Beatty (1976) Epidemics. New York.: Charles Scribner & Sons.

Meegan, J.M. (1979) The Rift Valley fever epizootic in Egypt, 1977-1978. 1. Description of the epizootic and virologic studies. Transactions of the Royal Society of Tropical Medicine and Hygiene 73(6):630-633.

Moore, R.M., Jr., R.B. Zelmer, J.I. Moulthrop, and R.L. Parker (1977) Surveillance of animal-bite cases in the United States, 1971-1972. Archives of Environmental Health 32(6):267-270.

National Institute of Mental Health (1972) Pollution: Its Impact on Mental Health, A Literature Survey and Review of Research. National Clearinghouse for Mental Health Information, Report No. 72-9135. Washington, D.C.: U.S. Department of Health, Education, and Welfare.

Neilsen, L.T. (1979) Mosquitoes, the mighty killers. National Geographic 156(3):426-440.

Nelson, B.C. (1978) Ecology of medically important arthropods in urban environments. Pages 87-124, Perspectives in Urban Entomology, edited by G.W. Frankie and C.S. Koehler. New York: Academic Press, Inc.

New York City Department of Health (1978) Tick Control Program in New York City. Annual Activities. Submitted to New York State Department of Health. (Unpublished)

Olkowski, W., H. Olkowski, A.I. Kaplan, and R. van den Bosch (1978) The potential for biological control in urban areas: shade tree insect pests. Pages 311-347, Perspectives in Urban Entomology, edited by G.W. Frankie and C.S. Koehler. New York: Academic Press, Inc.

Olkowski, W., L. Laub, and H. Olkowski (1979) The Natural Enemies of Shade Tree Pest Insects: A National Survey. Center for the Integration of the Applied Sciences, John Muir Institute, Berkeley, California. EPA Contract No. R804205-01. Washington, D.C.: U.S. Environmental Protection Agency.

Pal, R. (1962) Contributions of insecticides to public health in India. World Review of Pest Control 1(2):6-10.

Palika, P., L. Malis, and K. Zwyrtrek (1971) To the problem of scabies and pediculosis in Karvina region. Ceskoslovenska Hygiena, Epidemiologie, Mikrobiologie, Immunologie 12:54-59.

Pratt, H.D. and J.S. Wiseman (1962) Fleas of Public Health Importance and Their Control. Communicable Disease Center, U.S. Public Health Service. Atlanta: U.S. Department of Health, Education, and Welfare.

Pratt, H.D., B.F. Bjornson, and K.S. Littig (1976) Control of Domestic Rats and Mice. HEW Publication No. (CDC) 76-8141. Atlanta: U.S. Department of Health, Education, and Welfare.

Roth, L.M. and E.R. Willis (1957) The medical and veterinary importance of cockroaches. Smithsonian Miscellaneous Collection 134. Washington, D.C.: Smithsonian Institution.

Roth, L.M. and E.R. Willis (1960) The biotic association of cockroaches. Smithsonian Miscellaneous Collection. Vol. 141. Washington, D.C.: Smithsonian Institution.

Ruebush, T.K., P.B. Cassady, H.J. Marsh, S.A. Lisker, D.B. Voorhees, E.B. Mahoney, and G.R. Healy (1977a) Human babesiosis on Nantucket Island. Annals of Internal Medicine 8(1):6-9.

Ruebush, T.K., D.D. Juranek, E.S. Chisholm, P.C. Snow, G.R. Healy, and A.J. Sulzer (1977b) Human babesiosis on Nantucket Island: Evidence for self-limited and subclinical infections. New England Journal of Medicine 297:825-827.

Rueger, M.E. and T.A. Olson (1969) Cockroaches (*Blattaria*) as vectors of food poisoning and food infection organisms. Journal of Medical Entomology 6:185-189.

Ryan, J., R. Stoller, and L. Moore (1974) Rural and urban residents differ in their knowledge and attitudes about the use of insecticides. Progressive Agriculture in Arizona 26(3):6-7, 16.

Salcedo, R.N., H. Read, J.F. Evans, and A.C. King (1971) Rural-urban perspectives of pesticide industry. Rural Sociology 36:554-557.

Savage, E.P. (1978) National Household Pesticide Usage Study, 1976-1977. Colorado State University. EPA Contract No. 68-01-4663. (Unpublished draft)

Schantz, P.M. and L.T. Glickman (1979) Canine and human toxacariasis: The public health problem and the veterinarian's role in prevention. Journal of the American Veterinary Medical Association 175(12):1270-1273.

Scheone, H., F. Falah, and A. Rojas (1973) La infestacion por Pediculus humanus capitis en Santiago de Chile. Boletin Chileno de Parasitologia 28:31-33.

Shaw, P.K. and D.D. Juranek (1976) Recent trends in scabies in the United States. Journal of Infectious Diseases 134(4):414-416.

Stoller, A., J. Krupinski, A.J. Christophers, and G.K. Blanks (1965) Organophosphorus insecticides and major mental illness. Lancet 7400:1387-1388.

Tait, E.J. (1979) Measuring Attitudes to Risk: Farmers' Attitudes to the Financial, Personal and Environmental Risks Associated with Pesticide Usage. Cambridge, England: University of Cambridge. (Unpublished)

The White House (1979) President's Message on the Environment. Washington, D.C.: Office of the White House Press Secretary.

U.S. Department of Health, Education, and Welfare (1979) Asthma and Other Allergic Diseases. Report of a Task Force of the National Institute of Allergy and Infectious Diseases. U.S. Public Health Service, Publication No. 79-387. Bethesda, Md.: National Institutes of Health.

U.S. Public Health Service (1968-1977) Annual Surveillance Reports. Center for Disease Control. Atlanta: U.S. Department of Health, Education, and Welfare.

U.S. Public Health Service (1976) Recommendations of the Public Health Service Advisory Committee on Immunization Practices: Rabies Prophylaxsis. Center for Disease Control. Morbidity and Mortality Weekly Report 25(51):404-406.

U.S. Public Health Service (1979a) Reported morbidity and mortality in the United States, 1978. Center for Disease Control. Morbidity and Mortality Weekly Report, Annual Supplement 27(54):3.

U.S. Public Health Service (1979b) Rickettsial Disease Surveillance Report No. 1, 1975-1978. Center for Disease Control. Atlanta: U.S. Department of Health, Education, and Welfare.

U.S. Public Health Service (1979c) Malaria Surveillance Annual Summary, 1978. Center for Disease Control. Atlanta: U.S. Department of Health, Education, and Welfare.

U.S. Public Health Service (1980a) Conference reviews recent developments in dengue activity in North America. Center for Disease Control. Morbidity and Mortality Weekly Report 29(7):75.

U.S. Public Health Service (1980b) Rabies Surveillance, Annual Summary 1979. Center for Disease Control. Atlanta: U.S. Department of Health, Education, and Welfare. (In Press)

von Rumker, R., R.M. Matter, D.P. Clement, and F.K. Erickson (1972) The Use of Pesticides in Suburban Homes and Gardens and Their Impact on the Aquatic Environment. Pesticide Study Series 2. Office of Water Programs, U.S.

Environmental Protection Agency. Washington, D.C.: U.S. Government Printing Office.

Whyte, A.V.T. (1977) Guidelines for Field Studies in Environmental Perception. Paris: UNESCO.

Winkler, W.G. (1977) Human deaths induced by dog bites, United States, 1974-1975. Public Health Reports 92(5):425-429.

Winkler, W.G. and K.D. Kappus (1979) Human antirabies treatment in the United States, 1972. Public Health Reports 94(2):166-171.

World Health Organization (1976) Vital Statistics and Causes of Death. World Health Statistics 1973-1976. Vol. 1. Geneva: World Health Organization.

World Health Organization (1977) 22nd Report of Expert Committee on Insecticides. Technical Report Series, WHO No. 585. Geneva: World Health Organization.

Zinsser, H. (1935) Rats, Lice and History. Boston, Mass.: Little, Brown, & Company.

3

Management of Urban Pests

CHEMICAL MANAGEMENT METHODS

Survey of Chemical Controls

Information on the kinds and quantities of chemical pesticides applied in urban areas by various user groups is essential to the interpretation of epidemiological and monitoring data and to the evaluation and possible improvement of current urban pest management programs. The following section summarizes available studies of chemical pesticide use in urban areas.

Data from Previous Studies and Reports

In the last 10 years a number of investigators have studied the use of pesticides for nonagricultural purposes from various perspectives. Table 3.1 summarizes studies of nonfarm pesticide use that were reviewed for this report. The reports listed in Table 3.1 fall into three groups: reports dealing with nonfarm pesticide use on a nationwide basis; statewide surveys; and reports pertaining to selected urban areas.

Nationwide Studies Of the reports in this group, the study by Savage (1978) is the most extensive and detailed. The study was conducted by the Epidemiologic Pesticide Studies Center of Colorado State University under contract to the Office of Pesticide Programs, U.S. Environmental

Protection Agency (EPA), and was the first serious attempt to develop information on household pesticide use on a national basis. Out of a statistically designed nationwide sample of about 10,000 households, interviews were obtained from 8,254.

For the 14 pesticides (containing 18 different chemical active ingredients) mentioned most frequently during the interviews, Savage reports the quantities of formulated product(s) used or stored by householders. The active-ingredient content of these products is not given, however, and therefore the data cannot be translated into total quantities of active ingredients used. Pesticide use by pest control operators, public agencies, and commercial, industrial, and institutional users was not covered.

A second nationwide study is that conducted recently by the Economic Analysis Branch of EPA's Office of Pesticide Programs. This study estimates farm and nonfarm use of pesticides in the United States in 1979 (U.S. EPA 1979d). Use is estimated both in terms of user expenditures and in terms of quantities of active ingredients for farm and nonfarm use categories and for pesticide categories. Table 3.2 shows the estimated quantities of pesticides used for farm and nonfarm purposes in 1979.

The U.S. Department of Agriculture (USDA) conducts periodic surveys of pesticide use in the United States, the two most recent having been conducted in 1971 and 1976 (USDA 1974, 1978a). Table 3.3 shows total U.S. production, total imports and exports of pesticides, the net domestic supply, and farm and nonfarm uses of synthetic organic pesticides in 1976 as reported by the USDA (1978a, 1978b). Table 3.3 reflects one of the major problems with statistics of this sort, which is considerable uncertainty about the size of the total domestic supply of pesticides and, consequently, about the magnitude of nonfarm pesticide use.

In the USDA reports the figures for farm use were based on user surveys, while nonfarm use was derived by subtracting farm use from total U.S. supply. The U.S. supply, in turn, was derived by adding imports and subtracting exports from U.S. production. Variances between quantities reported as "U.S. supply" are due in part to differences in the statistical treatment of inorganic pesticides, fumigants, soil conditioners, wood preservatives, and others by different government agencies keeping data on production, imports, and exports. Another source of discrepancy is the lack of information on the active-ingredient content of pesticides that are exported.

Despite these difficulties, the data on nonfarm pesticide use in the United States as reported by the USDA for 1976 in Table 3.3 and by EPA for 1979 in Table 3.2 are reasonably compatible. Both sets of data were of limited value to the Committee on Urban Pest Management, however, because they were not broken down into narrower categories.

TABLE 3.1 Characteristics of Reports on Nonfarm Pesticide Use in the United States[a] Nationally and in Selected States and Cities

Characteristic	National					State			Local				
	Savage (1978)	U.S. EPA (1979d)	USDA (1974, 1978a)	Keil et al. (1977)	von Rumker et al. (1974)	California (1979a, b)	Kamble et al. (1978)	Colorado (1979)	Frankie & co-authors (1978, In press)[b]	Olkowski et al. (1978a)	Olkowski et al. (1978c)	Lande (1975)	von Rumker et al. (1972)
Area studied:													
United States	X	X	X	X	X	–	–	–	–	–	–	–	–
by regions	X	–	–	X	X	–	–	–	–	–	–	–	–
States	–	–	–	X	–	X	X	X	–	–	–	–	–
Selected urban areas[c]	X(25)	–	–	–	–	–	–	–	X(4)	X(5)	X[d]	X(1)	X(3)
User groups surveyed:													
Households	X	–	–	–	–	–	–	–	X	–	–	X	X
Pest control operators	–	–	–	X	X	X	X	X	X	–	–	–	–
Public agencies	–	–	–	X	X	X	–	–	–	X	X	X	X
Comm./ind. users	–	–	–	X	X	X	–	–	–	–	–	X	X

Pests surveyed:														
Indoor pests	–	–	–	–	X	X	–	X	X	–	X	X	–	X
Outdoor pests	–	–	–	–	X	X	–	X	X	X	X	X	–	X
Site(s) studied:														
Indoor	X	–	–	–	X	X	X	X	X	–	–	–	X	–
Outdoor	X	–	–	–	X	X	X	X	X	X	–	–	X	X
Nonfarm pesticide usage reported by:														
Categories[e]	X	X[g]	X[h]	–	X	X[j]	–	–	–	X	X	X	X	X[l]
Chemicals (AI)	X	–	–	–	X[i]	X[j]	X[m]	X	X	–	X	X	X	X[l]
by quantities	–[f]	–	–	–	X[i]	X[j]	X[m]	X	–[n]	–	X[k]	–	–	X[l]
Pesticide suppliers surveyed	X	–	–	–	X	X	–	–	–	–	–	–	–	X
Study year	1976-1977	1979	1971	1976	1974	1972	Annual	1978	1977-1978	1974-1976, 1977-1978	1976-1977	1977	1973	1971

[a]Explanation of symbols:

X = covered in report

− = not covered in report

[b]The studies on urban pest management by G. W. Frankie and co-authors available at this writing are as follows:

Frankie and Levenson (1978)

Levenson and Frankie (In press)

Frankie and Magowan (In press)

Granovsky and Frankie (In press)

[c]Number in parentheses following X-mark indicates number of urban areas covered in study.

[d]In surveys reported in this paper, an unspecified number of California county agricultural commissioners and farm advisers and 16 California cities responded to questionnaires by mail and/or by telephone.

[e]"Categories" = herbicides, insecticides, fungicides, etc.

[f]Due to funding limitations, collection of pesticide use data by quantities of active ingredients was not included in the scope of work of this study. For 14 pesticides observed most frequently, the author reports quantities of formulated product(s) used or stored by the households interviewed, but the active ingredient content of these products is not given.

[g]Nonfarm pesticide uses estimated for 1979 by EPA staff, broken down by categories (herbicides; insecticides; fungicides; and rodenticides, fumigants and molluscicides) and by industrial/commercial/governmental and home and garden uses, respectively, based on National Agricultural Chemicals Association annual surveys, U.S. International Trade Commission data, and other sources. See Table 3.2.

[h]Nonfarm pesticide uses, broken down by categories (fungicides; herbicides; and insecticides, miticides and fumigants), derived as difference between farm use and estimated total domestic use of pesticides. See Table 3.3.

[i]Nonfarm pesticide usage was estimated at only 6% of total U.S. usage in this study. The USDA (1974, 1978a, b) and U.S. EPA (1979d) estimate nonfarm uses at 35-40% and 27%, respectively, of total domestic pesticide usage.

[j]Due to funding constraints, the sponsoring agencies (EPA and CEQ) limited this study to 25 selected pesticides, and excluded home and garden pesticide uses from the proposed scope of work.

[k]Quantitative pesticide use data reported are "by no means representative," according to the authors.

[l]This study was limited to pesticide usage in suburban homes and gardens by the sponsoring agency's (EPA) scope of work.

[m]Includes only about 85% of total usage of pesticides requiring a permit in California, and much smaller, but unknown percentages of the usage of nonrestricted pesticides. Does not include nonrestricted pesticides used by private, unlicensed applicators.

[n]Pesticides used by pest control operators reported by quantities of formulations whose active ingredient content is not stated.

TABLE 3.2 Estimated Farm and Nonfarm Uses of Pesticides in the
United States in 1979, by Quantities[a]

Use	Insecticides[b]	Herbicides[c]	Fungicides[d]	Other[e]	Total
Farm uses	302,400	448,000	37,100	53,000	840,500
Nonfarm uses Industrial/ Commercial/ Government	37,800	84,000	58,300	42,400	222,500
Home & garden	37,800	28,000	10,600	10,600	87,000
Subtotal	75,600	112,000	68,900	53,000	309,500
Total:	378,000	560,000	106,000	106,000	1,150,000

[a] All quantities reported in thousands of pounds of active ingredients.
[b] Includes miticides and contact nematicides.
[c] Includes plant growth regulators.
[d] Does not include wood preservatives.
[e] Includes rodenticides, fumigants, and molluscicides.

SOURCE: U.S. EPA (1979d), based on National Agricultural Chemicals Association
annual surveys and U.S. International Trade Commission data.

Keil et al. (1977) studied the use of pesticides in the agricultural,
governmental, and industrial sectors in the United States in 1974. The
study was conducted by the Epidemiologic Studies Program Center at the
Medical University of South Carolina under an EPA contract. Household
pesticide use was not included. Estimates of farm pesticide use were
obtained from state pesticide coordinators, while information on industrial
pesticide use was obtained from public utilities, pest control operators, and
national distributors of pesticides. Data on pesticide use by governmental
agencies was obtained from state health departments, state highway
departments, park and forest agencies, military installations, several U.S.
Department of Agriculture agencies, the Army Corps of Engineers, the
U.S. Postal Service, and the U.S. Department of the Interior.

Keil et al. reported that 903.2 million lbs of active ingredients of all
types of pesticides were used in the United States in 1974. Of that total, 94
percent was used in agriculture, 3.5 percent was used by governmental
agencies, and 2.5 percent was used in the industrial sector. The finding that
nonfarm uses constituted only 6 percent of all pesticide uses in the United
States was greatly at variance with other reports, and the authors cited

four possible explanations for the discrepancy: overestimation of farm uses, underestimation of governmental and industrial uses, the failure to consider household uses, or mistakes in other studies. The large discrepancies between the results of this study and the findings of other investigators on the magnitude of nonfarm pesticide use, and the fact that household and home garden pesticide uses were not considered by Keil et al., limited the usefulness of the data for this report.

In 1972 von Rumker et al. (1974) studied the production, distribution, use, and environmental impact potential of selected pesticides. Because of funding constraints, the sponsoring agencies (EPA and the Council on Environmental Quality) limited the investigation to 25 pesticides and excluded household and home garden pesticide uses from the proposed scope of work. Where necessary to complete case studies of the 25 selected pesticides, the quantity used in households and home gardens was deemed to be the difference between U.S. production plus imports and other domestic uses plus exports. Because household uses were not explicitly considered, the study was of limited usefulness to the Committee.

TABLE 3.3 Production, Imports, Exports, Domestic Supply, and Farm and Nonfarm Uses of Synthetic Organic Pesticides in the United States in 1976, by Quantities[a]

Pesticides:	Insecticides[b]	Herbicides[c]	Fungicides[d]	Total
United States production	566,084	656,217	142,090	1,364,391
Imports	4,723	47,376	8,443	60,542
Exports	290,355	207,179	46,209	543,743
United States supply[e]	280,452	496,414	104,324	881,190
United States supply[f]	350,000	555,000	110,000	1,015,000
Farm uses[g]	208,000	410,000	43,000	661,000
Nonfarm uses	142,000	145,000	67,000	354,000

[a] All quantities reported in thousands of pounds of active ingredients.
[b] Includes miticides, nematicides, livestock insecticides, rodenticides, and organic fumigants.
[c] Includes plant growth regulators.
[d] Includes some wood preservatives. Does not include disinfectants.
[e] Derived by adding imports, and subtracting exports from production, based on data reported by USDA (1978b).
[f] Reported by USDA (1978a). Discrepancies between the two sets of data for "U.S. supply" are due to differences in treatment of certain pesticide categories (e.g., inorganic pesticides, fumigants, soil conditioners, wood preservatives, etc.), and to lack of information on the active ingredient content of exported pesticides.
[g] Does not include forestry uses.

SOURCE: U.S. Department of Agriculture (1978a, b).

State Studies A number of state agencies survey the use of pesticides. Many of these surveys focus on farm use, but some include nonfarm use. Information from reports on nonfarm use in California, Nebraska, and Colorado are included in Table 3.1. (A number of other states collect and publish similar information on various nonfarm pesticide uses.)

Of all the states, California maintains the most complete and detailed reporting system of pesticide uses (California Department of Food and Agriculture 1979a, 1979b, and earlier annual Pesticide Use reports). Data on pesticide use in California are compiled and summarized both quarterly and annually from reports from structural and agricultural pest control operators, public agencies, and growers applying restricted pesticides. However, the reports reflect only a portion of total pesticide use in the state; the California Department of Food and Agriculture states that the reports reflect approximately 85 percent of the total use of pesticides whose application requires a permit under the California Administrative Code. For nonrestricted pesticides, the reports include only quantities applied by licensed pest control operators. Nonrestricted pesticides used by private, unlicensed applicators are not included. Since a large portion of the pesticides used in urban and suburban households, home gardens, and commercial and industrial establishments is nonrestricted, even the California reports do not provide definitive information on urban or suburban pesticide use in the state.

Kamble et al. (1978) examined the types, amounts, and formulations of pesticides used by the structural pest control industry in Nebraska in 1978. The study was funded in part by the Pesticide Branch of the EPA Region VII office and was carried out by the Institute of Agriculture and Natural Resources, University of Nebraska at Lincoln. All 52 firms identified as being involved in structural pest control in Nebraska responded to a letter questionnaire followed by telephone calls and personal visits. The survey indicated that respondents used 100 pesticide formulations with 49 different active ingredients. The results were reported in terms of individual pesticides, quantities of active ingredients, application methods, and target pests.

Although the report presents a detailed account of pesticide use by the 52 structural pest control firms in Nebraska, other urban and suburban pesticide uses were not examined. Consequently, the results do not reflect and cannot be extrapolated to show total urban and suburban pesticide use in Nebraska.

According to Kamble (University of Nebraska, personal communication, 1979), additional studies recently conducted in Nebraska included a survey of pesticide use by homeowners. The field survey work has been completed, but the results were not available for this report.

The Colorado Department of Agriculture (1979) collected data on pesticide use by structural pest control operators during the period July 1, 1977 to June 30, 1978. Pesticides used by 55 licensed pest control firms in Colorado were reported by amounts of various formulations, but the active ingredients were not stated. Colorado does not collect data on pesticides used in urban areas by nonlicensed applicators such as homeowners and building maintenance personnel (R.B. Turner, Colorado Department of Agriculture, personal communication, 1979).

Local Studies Of the studies in Table 3.1 involving pesticide use in selected urban areas, the investigations by Frankie and co-authors are the most recent and most comprehensive. The results are being reported in a series of publications, of which four were available at the time of the Committee's work: Frankie and Levenson (1978), Levenson and Frankie (in press), Frankie and Magowan (in press), and Granovsky and Frankie (in press).

The investigators studied public opinion, attitudes, and pest control practices in three urban areas (Berkeley, California; Dallas, Texas; and New Brunswick, New Jersey), following an exploratory study in College Station, Texas, and Dallas. The surveys focused on households and pest control operators and did not include other urban pesticide users, such as public agencies and commercial, industrial, and institutional users. Pesticide use was reported for individual pesticides used to control various indoor and outdoor pests, primarily arthropod pests.

Olkowski et al. (1978b) studied the management of pests on street trees in five California cities: Berkeley, San Jose, Palo Alto, Modesto, and Davis. The number of treatments with insecticides was reported, but the study is not explicit about the active ingredients or the quantities of pesticides used.

In another study, Olkowski et al. (1978c) obtained information on urban pest management problems and practices by interviewing an unspecified number of California county agricultural commissioners and farm advisers in 16 California communities. Target pests, pesticide categories, and the active ingredients and quantities of individual pesticides were reported. The authors emphasize, however, that the data are "by no means representative."

In 1971 von Rumker et al. (1972) studied the use of pesticides in suburban homes and gardens in three selected metropolitan areas: Philadelphia; Dallas; and Lansing, Michigan. The study included all pesticide user groups (households, pest control operators, public agencies, and commercial/industrial users). Pesticide use was reported by categories, chemical active ingredients, and quantities. However, in accordance with the sponsoring agency's (EPA) scope of work, the study was limited to suburban areas and did not include pesticide use in inner cities. The authors estimated annual per capita use at 0.14 lb. In 1971, the 5,465,000 suburban residents of these three cities applied an estimated 130,000 lbs of herbicides, 510,000 lbs of insecticides, and 120,000 lbs of fungicides. On an area basis homeowners applied from 5.3 to 10.6 lbs of pesticide per acre, suggesting that suburban lawns and gardens receive heavier pesticide applications than most other land areas in the United States.

Lande (1975) made a random survey of pesticide use practices in Allegheny County, Pennsylvania, in 1973. The groups surveyed included households, public agencies, and commerical, industrial, and institutional users, and results were reported for frequency of use of types of pesticides and individual products at various sites (primarily outdoor). Quantities of active ingredients were not reported.

Data from EPA and Other Sources

As noted above, most of the studies examined provide limited, and often deficient, information about urban uses of pesticides. Because of the lack of adequate quantitative data on urban pesticide use in the reports reviewed, the Committee asked EPA for information about:

1. Chemical pesticides registered by EPA for control or management of urban pests;
2. Volume of production and use of the chemical pesticides for urban pest management broken down by products, quantities, and sites of use; and
3. Geographical distribution of use.

Following receipt of the request, EPA asked the Committee to select from EPA's computerized list of pesticide use sites and site categories those in which the Committee was interested. Also, EPA advised that it would be very costly and time consuming to generate through its computer a list of all pesticide chemicals registered for use in the sites to be specified by the Committee; EPA requested instead that the Committee furnish a list of the pesticides in which it was interested. Such a list was compiled by the Committee staff in consultation with several Committee members. The list included 9 insecticides, 3 fungicides, 2 herbicides, and 13 rodenticides (see Tables 3.7, 3.8, and 3.9). These 27 pesticides were made the subject of a formal, internal-Agency request for computer services dated July 23, 1979. By the Committee's report-writing deadline 1 month later, EPA had processed 5 of 9 insecticides, 2 of 3 fungicides, 1 of 2 herbicides, and most of the rodenticides on the Committee's list. Information on these products was delivered to the Committee not by way of a few simple tables as anticipated, but in the form of approximately 6,500 full-size computer-printed pages approximately 2 1/2 ft (76 cm) in height.

That information was produced from the EPA Office of Pesticide Programs' computerized data system on federally registered pesticide products. The system includes information on pesticide active ingredients and formulated products, product names, manufacturers, residue tolerances, and site and pest information for active ingredients and products. These data are also available on microfiche; the complete set of pesticide product information on microfiche consists of approximately 370,000 full-size computer pages reduced to about 1,400 microfiche cards (270 computer pages per card).

The information received from EPA provided partial answers to the Committee's first two questions for 18 of the 27 specified pesticides, but did not include any quantitative data on volume of production, or use by chemicals, products, sites, geographical distribution, or information on use restrictions. The amazing bulk of EPA's reply was due to the fact that it consists mostly of numerous repetitions of the same pesticide active ingredients, products, and combination products registered for control of each pest in each site. The cost of producing this information was approximately $2,000.

After preliminary examination of the data, the Committee wished to learn more about the computerized pesticide data system from which they were generated, and therefore examined a complete set of EPA's "pesticide product information on microfiche." This resulted in the question of whether the Committee's information needs could have been served more efficiently and simply by a computer run of the previously specified urban site categories vs. chemical active ingredients registered for use in these sites. The Agency did not respond to this question.

Therefore, an effort was made to manually extract information from the 6,500 computer-printed pages as well as from the data on microfiche. Consideration of the EPA information, supplemented by information from other sources, led to conclusions about target pests, the sites of pesticide use, and frequency of use for pesticides used in urban pest management. The results are presented in terms of pesticide categories and individual pesticide chemicals, but amounts of active ingredients are not quantified because of lack of information.

Target Pests Table 3.4 includes a listing of rodenticides and the rodent and other vertebrate pests against which they are registered and used in urban pest management.

The National Pest Control Association (NPCA) furnished the Committee a tabular summary of its Insect Control Committee's recommendations (NPCA 1979d) on insecticides for various pests (see Table 3.5). This list was sought after it became clear that preparing such a list from the EPA computer printouts and microfiche data would be complicated and time consuming. The Committee feels that the NPCA list provides a reasonably comprehensive and accurate description of insect and other arthropod pests associated with structures, and the pesticides used for their control.

Target Sites   EPA's pesticide information system includes a computerized list of all the types of sites for which pesticide chemicals are registered for use. There are 71 major site categories, including 32 describing agricultural sites—crops, animals, premises, and other areas; 7 describing miscellaneous nonagricultural sites; and 32 describing sites that may involve urban pesticide uses. It is the last group that was of interest to the Committee.

Table 3.6 shows the 14 urban use sites for which 9 insecticides, 2 herbicides, and 2 fungicides selected by the Committee are registered. The table also includes the number of EPA-registered products containing the specified chemicals as active ingredients. PCP, used primarily as a wood preservative, was omitted from Table 3.6 because its pattern of use is distinctly different from that of other pesticides commonly used in urban areas.

Table 3.4 shows similar information for the 14 rodenticides.

Frequency of Pesticide Use   Tables 3.7, 3.8, and 3.9 list insecticides and molluscicides, herbicides and fungicides, and rodenticides, respectively, used in urban pest management, by chemical and functional categories, common and trade names, and relative importance to users or frequency of use as perceived or reported by several groups: members of the Committee on Urban Pest Management (CUPM); members of urban households (Savage 1978); and pest control operators (NPCA 1979a, c, d, e). (Pest control operators are not included in Table 3.8, which lists herbicides and fungicides, because control of weeds and plant diseases is not usually a major part of their work.) The tables show that a number of pesticides were considered important or were used frequently by all three groups, while other products were considered important by only one or two groups. Pesticides frequently observed in urban households by Savage (1978) or rated important by pest control operators are listed in Table 3.10.

Von Rumker et al. (1972) pointed out that literally hundreds of people at the federal, state, and local levels were engaged in collecting information on the use of agricultural pesticides. By contrast, very little attention has

TABLE 3.4 Rodenticides Used in Urban Pest Management, by Number of Registered Products, Target Pests, and Use Sites[a]

Rodenticide[b] (Common Name)	Number of Registered Products	Target Pests	Use Sites								
			Domestic Dwellings	Sewage Systems	Uncultivated Nonfarm Areas	Wide Area Treatments	Recreational Vehicles	Eating Establishments	Food-Handling Facilities	Comm., Ind., Instit. Premises	Garbage Dumps
Warfarin	143	Rats, mice	X	X	–	–	X	X	X	X	X
Coumafuryl	60	Rats, mice	X	X	–	–	–	X	X	X	X
Chlorophacinone	13	Rats, mice	X	X	–	–	–	–	X	X	–
Diphacinone	73	Rats, mice, pigeons	X	X	–	X	–	–	X	X	X
Pindone	66	Rats, mice, pigeons, squirrels	X	–	–	–	–	–	X	X	–
PMP	3	Rats, mice	X	–	–	–	–	–	X	X	–
Pyrinuron	2	Rats, mice	X	–	–	–	–	–	X	X	X

Na-fluoroacetate	2	Rats, other rodents	—	—	—	X	—	X	X	X
Fluoroacetamide	1	Rats, other rodents	—	X	—	—	—	—	—	—
Antu	5	Rats	X	—	—	—	X	X	—	X
Zinc phosphide	12	Rats, mice, other rodents	X	—	—	—	X	X	X	X
Arsenic trioxide	2	Mice	X	—	—	—	X	X	X	—
Strychnine	5	Mice, pest birds, spiders, insects	X	—	—	—	X	X	X	—
Red squill	15	Rats	X	—	—	X	X	X	X	X

[a] Explanation of symbols:
X = chemical registered by EPA for use in specified site.
— = chemical not registered by EPA for use in specified site.

[b] Trade names and other names for these products are included in Table 3.9.

According to EPA's user guide for the information on microfiche, the pest/chemical, chemical/site/pest, and site/pest/chemical/registration number data sets are 70% accurate and complete. (In Committee work with these sets, several inconsistencies were noted between the microfiche and the computer-printed data.) EPA advises users of these data not to take any "substantive action" without prior data verification with source documents.

SOURCE: Retrieved manually from U.S. EPA (1979a, b).

TABLE 3.5 Chemical Controls for Insects and Related Pests Recommended by National Pest Control Association

Insects & Related Pests	Baygon® Residual Spray[c]	Baygon® Bait[c]	Baytex® Residual Spray[c]	Baytex® Granules[d]	Boric Acid Dust[c]	Cygon®[d]	DDVP Contact[b]	DDVP Space[b]	DDVP Resin Strip[d]	DDVP Bait[c]	Diazinon Residual Spray[c]	Diazinon Dust[c]	Diazinon Encapsulated[c]	Diazinon Granules[d]	Dibrom® Contact[b]	Dipterex® Bait[c]	Drione® Dust[c]	Dursban® Residual Spray[c]	Dursban® Granules[d]	Ficam® D Dust[c]	Ficam® W[d]	Gardona® Residual Spray[b]	Kelthane® Residual Spray[b]	Malathion Residual Spray[b]	Malathion Dust[c]	Malathion Bait[c]	Perthane®[c]	Pyrethrins Space[a]	Pyrethrins Contact[a]	Pyrethrins Dust[a]	Romel Residual Spray[b]	SBP-1382® Contact[b]	SBP-1382® Space[b]	Sevin® Dust[c]	Sevin® Residual Spray[b]	Sevin® Granule[d]	Sodium Fluoride Dust[c]
Ant	1	2	2	2			2	2			1	1	1	2	2			1	2	1	1			2	2		2	2	2	2	2	2		2	2	2	2
Bedbug								2									2											2	1	1	1						
Bee	1														1			1		1	1			2				2	2					2	2		
Book lice											2																										
Box elder bug							2	2			1	2						1						2	2			2	1								
Carpet beetle	2						2	2			1							1						2	2		1	2	2	2	2	2					
Centipede	1						2	2												2	1							1	1	1	1	2	2	2			
Clothes moth																											1	1	1	1	1			2			
Clover mite	2						2	2			1				2			1					1	2				2	1		1	2	1				
Cockroach	1	1	2	2			2	2			1	2	1	1	1			1	1	1	1			2	2		1	1	1	2	1	1			2		2

Pest																							
Cricket	1	2	1				2	2			2	2	2	1			2	2		2	2	2	2
Earwig	1			2	2		2	1			2	1	2	1			2	2	2	1		2	2
Fleas, cat/dog	2	2	2	2	1		2	2			2	2	1	1			2	2	2	1	2	2	2
Flies	2	1	1	2	1	1		1	1		1	1		1	1	2	2	1	1	2	1		
Grain/cereal beetle	1		2	2	2						1		1	1			1	2	2	1		1	
Millipede	1		2	1		2		1	2	2	1		2			2	2	2	2		2	2	2
Mosquitoes	1	1	1	1	2			1	2		1	1				1	1		1				
Scorpion	1		1			1		1			2	2				2	2	2	2				
Silverfish/firebrat	2	1	1	1	2	1	2	1			1	1	1	1		1	2	2	1	2	2	1	
Sowbug/pillbug	2	2		2		1			2	2	1	1				2	2	2	2	2		2	
Spider	1		2	1	2	1		1			1	1	1			2	2	1	1	2	2	1	1
Springtail	1			1			2				2					2							
Tick	1		2	2	1		2	1	2	2	1	1				2	2	2	2	2	2	2	2
Wasp	1	1	1	2			1	1	1		1	1				2	1	2	2	1	2	2	2

Key to numbers:
Recommended insecticides have been separated into two groups—1: Material of choice of proven value to PCO's in controlling this pest. 2: Less effective though some formulations may be useful. No entry: Pest not listed on label.

[a] Short residual (less than one day) for quick knockdown.
[b] Moderate residual (1-15 days), general spray.
[c] Long residual (more than 15 days), limited to crack and crevice or spot treatments.
[d] Long residual (more than 15 days), general spray.

SOURCE: NPCA (1979d).

TABLE 3.6 Insecticides, Herbicides and Fungicides Used in Urban Pest Management, by Use Sites and Number of Registered Products[a,b]

Pesticide use sites	Insecticides									Herbicides		Fungicides	
	Pyrethrins[c]	Malathion[d]	Diazinon[d]	Chlordane	Chlorpyrifos	Propoxur	Dichlorvos	Methoxychlor	Disulfoton[d]	2,4-D[e]	Paraquat	Captan	PMA
Ornamental plants, shrubs, and trees	X	X	X	X	X	X	X	X	X	—	X	X	—
Lawns and turf	X	X	X	X	X	X	X	X	X	X	X	X	X
Nonfarm soil treatments	X	X	X	—	—	—	—	—	X	—	—	—	—
Stored products	X	X	—	—	—	—	—	—	—	—	—	—	—
Pets[f]	X	X	X	X	X	X	X	X	—	—	—	X	—
Households (domestic dwellings)	X	X	X	X	X	X	X	X	—	X	—	X	—
Wood/wood structure protection	X	—	—	X	X	X	—	—	—	—	—	—	—
Aquatic sites	X	X	X	X	—	—	X	X	—	X	—	—	—
Uncultivated nonfarm areas	X	X	X	X	X	X	X	X	—	X	X	—	—
Wide area treatments (public health)	X	X	X	X	X	X	X	X	—	X	—	—	—
Antifouling treatments	—	—	—	—	—	—	—	—	—	—	—	—	—

Commercial, industrial, institutional uses	X	X	X	–	X	X	X	X	X	–	–	X	X	–
Domestic and human uses	X	X	X	–	X	X	X	X	X	–	–	–	X	–
Refuse and solid-waste sites	X	X	X	X	X	X	X	X	X	–	–	–	–	–
Number of EPA-registered products	305	210	154	90	71	54	31	16	24	61	11	32	3	

[a]Pesticides are listed by common names. (Trade names and other names of these products are included in Tables 3.7 and 3.8.)

[b]Explanation of symbols:

X = chemical registered by EPA for use in specified site.

– = chemical not registered by EPA for use in specified site.

[c]Product count and registration status by sites based on pyrethrin-piperonyl butoxide combination products.

[d]Product count does not include combination products.

[e]Product count includes only products containing 2,4-D dimethylamine salt as the sole active ingredient.

[f]Includes animal premises, pens, kennels, shelters, cages, transportation vehicles, bedding, equipment, manure treatment, etc.

According to EPA's user guide for the information on microfiche, the pest/chemical, chemical/site/pest, and site/pest/chemical/registration number data sets are 70% accurate and complete. (In Committee work with these sets, several inconsistencies were noted between the microfiche and the computer-printed data.) EPA advises users of these data not to take any "substantive action" without prior data verification with source documents.

SOURCE: Retrieved manually from U.S. EPA (1979a,b).

TABLE 3.7   Insecticides and Molluscicides Used in Urban Pest Management, by Chemical Class, Common and Trade Names, and Relative Importance to Users[a]

Chemical Class/ Common Name(s)	Trade Name(s)	CUPM Members[b]	House- holds[c]	Pest Control Operators[d]
Organophosphates				
Diazinon	Spectracide®, others	X	X	X (1)
Dichlorvos, DDVP	Vapona®, others	X	X	X (2)
Malathion	Many	X	X	X (2)
Chlorpyrifos	Dursban®	X	–	X (1)
Disulfoton	Di-Syston®	X	–	–
Dimethoate	Cygon®, others	–	–	X (3)
Fenthion	Baytex®, others	–	–	X (3)
Naled	Dibrom®	–	–	X (NR)
Trichlorfon	Dipterex®, others	–	–	X (3)
Tetrachlorvinphos	Gardona®	–	–	X (NR)
Ronnel	Korlan®, others	–	–	X (3)
Carbamates				
Propoxur, arprocarb	Baygon®	X	X	X (1)
Carbaryl	Sevin®	–	X	X (2)
Bendiocarb	Ficam®	–	–	X (1)
Chlorinated hydrocarbons				
Chlordane	Many	X	X	X (1)
Methoxychlor	Many	X	X	–
Lindane	Many	–	X	X (2)
Aldrin	Several	–	–	X (2)
Heptachlor	Several	–	–	X (2)
Ethylan	Perthane®	–	–	X (NR)
Dicofol	Kelthane®	–	–	X (NR)
Ovex[e]	Ovotran®, others	–	X	–
Other products				
Pyrethrins	Many	X	X	X (1)
Resmethrin	SBP-1382®, others	–	X	X (2)
Rotenone, rotenoids	Several	–	X	X (3)
Boric acid	Several	–	–	X (2)
Silica aerogel comb.	Drione®	–	–	X (2)
Sodium fluoride	Florocid®	–	–	X (3)
Molluscicide				
Metaldehyde	Several	–	X	–

[a]Explanation of symbols (see also note d below):

X = product considered important and/or used frequently by group specified.

– = product not considered important or not used frequently by group specified.

[b]Nine insecticides were considered of major importance in urban pest management by members of the NRC Committee on Urban Pest Management (CUPM).

[c]Twelve insecticides/miticides and one molluscicide were among the 18 pesticide

TABLE 3.7 Continued

chemicals used most frequently by 8,254 households interviewed in a nationwide household pesticide usage study (Savage 1978) conducted in the United States in 1976-1977.

[d]Twenty-five insecticides/miticides were reported used by pest control operators (PCO) in the United States. The relative importance of these products was rated by them as follows:

(1) Of primary importance in PCO work.

(2) Specific but minor applications in PCO work, or products less effective than the chemicals of choice.

(3) Only of occasional minor use.

(NR) Not rated.

[e]Product discontinued in United States.

SOURCE: NRC Committee on Urban Pest Management; Savage (1978); National Pest Control Association (1979c,d).

TABLE 3.8 Herbicides and Fungicides Used in Urban Pest Management, by Common Names, Trade Names, and Relative Importance to Users[a]

| Common Name(s) | Trade Name(s) | Importance/Frequency of Use as Perceived or Reported by: | |
		CUPM Members[b]	Households[c]
Herbicides			
2,4-D	Many	X	X
Silvex	Many	–	X
Paraquat	Paraquat®	X	–
Fungicides			
Captan	Orthocide®, others	X	X
Folpet	Phaltan®, others	–	X
PMA (phenylmercury acetate)	Many	X	–
PCP (penta-chlorophenol)	Many	X	–

[a]Explanation of symbols:

X = product considered important and/or used frequently by group specified.

– = product not considered important or not used frequently by group specified.

[b]Two herbicides and three fungicides were considered of major importance in urban pest management by members of the NRC Committee on Urban Pest Management (CUPM).

[c]Two herbicides and two fungicides were among the 18 pesticide chemicals used most frequently by 8,254 households interviewed in a nationwide household pesticide usage study (Savage 1978) conducted in the U.S. in 1976-1977.

SOURCE: NRC Committee on Urban Pest Management; Savage (1978).

TABLE 3.9   Rodenticides[a] Used in Urban Pest Management, by Mode of Action, Chemical Class, Common and Trade Names, and Relative Importance to Users[b]

Common Name(s)	Trade Name(s)	CUPM Members[c]	House- holds[d]	Pest Control Operators[e]
Multidose toxicants (anticoagulants)				
Coumarins				
Warfarin	Several	X	X	X (1)
Warfarin + sulfaquinoxaline	Prolin®	X	–	X (NR)
Coumafuryl	Fumarin®, Fumasol®	X	–	X (2)
Indandiones				
Chlorophacinone	Rozol®, Drat®	X	–	X (1)
Diphacinone	Diphacin®, Ramik®	X	–	X (2)
Pindone	Pival®	X	–	X (2)
PMP	Valone®	X	–	X (2)
Single-dose toxicants				
Synthetic organic chemicals				
Pyriminil, pyrinuron[f]	Vacor®, DLP-87 (787)®	X	–	X (NR)
Sodium fluoroacetate	1080®	X	–	X (3)
Fluoroacetamide	1081®	X	–	X (3)
α-naphthylthiourea, antu	Several	–	–	X (NR)
Inorganic chemicals				
Zinc phosphide	Phosvin®, others	X	–	X (1)
Phosphorus yellow[g]		X	–	–
Arsenic trioxide (As₂O₃)[g]	Several	–	–	–
Sodium arsenite[g]	Several	–	–	–
Botanicals				
Strychnine	Several	X	–	X (3)
Red squill	Several	–	–	X (3)

[a]Not including fumigants.
[b]Explanation of symbols (see also Note e below):
  X = product considered important and/or used frequently by group specified.
  – = product not considered important or not used frequently by group specified.

TABLE 3.9 Continued

[c]Thirteen rodenticides were considered of importance in urban pest management by members of the NRC Committee on Urban Pest Management (CUPM).

[d]One rodenticide, warfarin, was among the 18 pesticide chemicals used most frequently by 8,254 households interviewed in a nationwide household pesticide usage study (Savage 1978) in the United States in 1976-1977.

[e]Fourteen rodenticides were reported used by pest control operators (PCOs) in the United States. The relative importance of these products was rated by them as follows:

(1) Of primary importance in PCO work.

(2) Specific but minor applications in PCO work, or products less effective than the chemicals of choice.

(3) Only of occasional minor use.

(NR) Not rated.

[f]Voluntarily withdrawn by manufacturer.

[g]These products were EPA-registered as rodenticides in 1978, but were not reported being used by householders or pest control operators in recent surveys.

SOURCE: NRC Committee on Urban Pest Management; U.S. EPA (1978, 1979a,b,c); Savage (1978); National Pest Control Association (1979a,d,e).

been paid to the quantitative aspects of nonfarm pesticide uses and to the fate and disposition of pesticides applied for nonagricultural purposes. (Similar findings are recorded in most of the other reports included in Table 3.1.)

Moreover, at this writing, 8 years and a number of taxpayer-funded studies later, the situation remains basically unchanged. None of the studies available for this review and summarized in Table 3.1 provide data on the use of pesticides in American cities, broken down by chemical active ingredients and quantities, and none cover all four major urban user groups (households, pest control operators, public agencies, and commercial/industrial users). Most of the studies in selected urban areas were designed and conducted primarily for purposes other than the collection of pesticide use information, and the pesticide use data reported are incomplete and not suitable for extrapolation to other regions, let alone to the entire country.

In summary, it is our conclusion that the need for detailed quantitative data on the use of pesticides in urban areas of the United States, a need emphasized by most of the investigators whose studies are reviewed in this report, remains unfulfilled.

HEALTH CONCERNS WITH CHEMICAL MANAGEMENT OF URBAN PESTS

A variety of illnesses have been associated with exposure to chemical pesticides, and their clinical manifestations may vary with different types

TABLE 3.10 Pesticides Frequently Observed in Urban Households and/or Rated Important by Pest Control Operators

Pesticide (Common Name)	Chemical Class
Insecticides (see Table 3.7)	
Diazinon Dichlorvos (DDVP) Malathion Chlorpyrifos	Organophosphates
Propoxur Carbaryl	Carbamates
Methoxychlor Lindane Ovex	Chlorinated Hydrocarbons
Pyrethrins Resmethrin Rotenone	
Molluscicide (see Table 3.7)	
Metaldehyde	
Herbicides (see Table 3.8)	
2,4-D Silvex	Phenoxyacetic acid derivatives
Fungicides (see Table 3.8)	
Captan Folpet	Phthalimids
Rodenticides (see Table 3.9)	
Warfarin Chlorophacinone Zinc phosphide	

SOURCE: Tables 3.7, 3.8, and 3.9 of this report.

of exposure. Assessment of chemical pest control in an urban setting must therefore take into consideration what is currently known about human exposure so as to understand the associated adverse health effects of populations at risk.

Characterization of Pesticide Exposure

The principal routes of human exposure to pesticides in an urban setting include direct use of chemicals by householders using over-the-counter products, such as insect sprays and baits, and direct exposure to toxicants used by certified (commercial or government) applicators. For both householders and certified applicators the pesticide label constitutes the chief source of instruction and of use restrictions, although certified applicators have also undergone both formal and informal training.

Labeling, which is regulated by the U.S. Environmental Protection Agency, provides directions for recommended use. If it is determined that use of a pesticide, even in accordance with directions, may generally cause unreasonable adverse effects on human health or the environment, use is restricted and the product is so labeled. Only certified applicators are authorized to use or supervise the use of pesticides registered for restricted use only. All other products may be sold to the general public.

Pesticides can be absorbed through the mouth, the lungs, or the skin. Figure 3.1 illustrates the major types of human exposure to pesticides and the adverse health effects that have been linked with them. These observed health impacts have been sustained by agricultural workers and others who manufacture or apply pesticides. In agriculture the organophosphate and carbamate pesticides have been the major causes of acute systemic illness, most of which have been the result of dermal exposure. Acute poisoning has occurred because of misuse as well as overuse, usually where little or no safety training was given to the applicators and mixers, the occupational group most frequently poisoned.

Federal regulations have generally reduced the burden of acute pesticide toxicity. EPA requires the certification of applicators of restricted-use pesticides, a requirement that has resulted in a nationwide training program on pesticide safety. The training program and the subsequent certification of applicators are intended to considerably reduce the misuse and overuse of pesticides. A second regulation related to the certification program is the classification of pesticides into either restricted- or general- (nonrestricted) use categories. This classification scheme has focused on 50 or more active ingredients and the emphasis has been on agricultural use rather than on urban use.

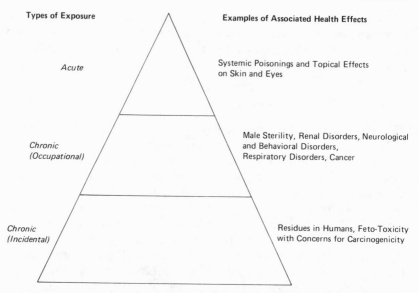

Types of Exposure

Acute

*Chronic
(Occupational)*

*Chronic
(Incidental)*

Examples of Associated Health Effects

Systemic Poisonings and Topical Effects
on Skin and Eyes

Male Sterility, Renal Disorders, Neurological
and Behavioral Disorders,
Respiratory Disorders, Cancer

Residues in Humans, Feto-Toxicity
with Concerns for Carcinogenicity

FIGURE 3.1   Types of pesticide exposure with examples of related health effects.

Chronic exposure is primarily occupational in nature and may also occur during the manufacture, storage, formulation, mixing, application, and disposal of pesticides; here, dermal and respiratory absorption are the major routes of exposure. Epidemiological recognition of occupational morbidity is facilitated wherever exposure is to a single pesticide. Neurological and behavioral problems have been recognized in workers manufacturing chlordecone (Sanborn et al. 1979) and leptophos (U.S. Congress, Senate 1976), and male sterility has been recognized in workers applying dibromo-chloropropane (DBCP) (Whorton et al. 1977).

In the urban setting it is the professional pest control operator who is at greatest risk of chronic exposure to pesticides. According to an NPCA survey, there are approximately 30,000 professional pest control operators in the United States. Of these, 69.3 percent are primarily concerned with general pest control and 27 percent with termite control (P. Spear and W. Jackson, Committee on Urban Pest Management, personal communication, 1979).

Exposure of the public at large is predominantly incidental, resulting primarily from residues in food, the home, clothing, dust, air, and water. Absorption of traces of pesticides occurs via all three routes (mouth, lungs, skin) mentioned earlier.

The exposure situation of the city dweller is quite different from that of the agricultural worker. In urban settings, pesticides are applied largely by untrained individuals with limited understanding of safety requirements, and to a lesser extent by trained pest control operators. Children are very much a part of this urban ecosystem and are at special risk of pesticide exposure (Reich et al. 1968). The variety of chemicals in use (see section on Survey of Chemical Controls) means that exposure is multiple, and the general population in the urban environment is at risk of exposure to chemicals that can produce either acute or incidental effects; the pest control operator has the additional risk of occupational exposure.

Pesticide Monitoring Programs

Because of inadequate quantitative information on pesticide use, it is difficult to measure human exposure in urban areas accurately. Accordingly, exposure assessment must rely heavily on information from monitoring programs (Yobs 1971). EPA's National Human Monitoring Program for Pesticides measures, among other things, incidental exposure of the general U.S. population to pesticide residues in food, house dust, soil, air, and water (Kutz et al. 1977). It is assumed that most of the body burden is acquired from residues resulting from urban pest management programs.

Exposure to newer pesticides has been identified by the monitoring program as exposure to older pesticides, such as DDT, has declined. This is a result of changes in use patterns, the development of newer chemical analytical methodologies, and mounting concern for risks to human beings and the environment. The multiresidue analytical techniques now being used for measuring urinary alkyl phosphates and phenolic metabolites provide an ideal instrument for present and future assessment of the chronic exposure of pesticide applicators and the safety potential of protective devices (Shafik et al. 1973a, b).

Definitive epidemiological evidence for the general urban population remains difficult to obtain, however, in the case of the newer pesticides. When the epidemiology of DDT residues in human beings in the United States was first studied, it was found that the residues were associated with social class, were higher in blacks than in whites, were higher in southern than in northern areas, and exhibited clustering by income (Davies et al. 1972). DDT residues in house dust demonstrated that urban pest management programs were a primary cause of exposure (Davies and Edmundson 1972, Davies et al. 1975). Studies of the residues of other organochlorine pesticides have not demonstrated differences as large as those seen with DDT, and most of the exposure can be explained by pesticide residues in food. At present, the most that can be said is that incidental pesticide exposure in the home is acquired from a variety of sources, and that the relative contribution of residues from food, dust, and urban pest control programs is different for different pesticides (Davies 1973, U.S. DHEW 1969).

Table 3.11 summarizes EPA data on organochlorine, alkyl phosphate, and phenolic metabolites in human beings (Kutz 1978, Kutz et al. 1978). The existence of many of the pesticide metabolites identified in urban settings probably reflects domestic exposure to specific organophosphate and carbamate insecticides. The data are useful indicators of subtle exposures to the newer generation of pesticides commonly used in urban pest management practices.

Acute Poisoning of Pest Control Operators

There is very little comprehensive information on pesticide exposure or poisoning among professional pest control operators (PCO) in the United States. California, with more than 2,000 licensed PCOs, probably has the best occupational injury reporting and record-keeping system in the country. In 1977, 1,531 cases of pesticide-related illness were reported by California physicians to the State Department of Food and Agriculture; 739 persons, including 24 pest control operators and 25 fumigators, developed systemic poisoning; the other cases of illness occurred in agricultural workers (Maddy 1978).

More recently, because of such pesticide-related health effects as cancer and male sterility, concern about frequent pesticide exposure as an occupational hazard has increased, and occupational health and safety strategies have been designed to minimize occupational exposure. The emphasis is on training, education, and certification of the applicator, improvements in protective clothing, adequate laundering of clothing, the use of new pesticide formulations, and safety improvements during the mixing and application of chemicals. Greater use of closed-circuit loading from mixing tank to aircraft to minimize exposure, and improved container disposal are examples of some of the modern techniques that have been developed to promote worker safety.

Direct Health Effects—Pesticide Poisoning Reports

The only proven effect of direct exposure to chemical pesticides in urban areas is acute poisoning, but even for poisoning, accurate and verifiable statistics are not available for most areas of the United States. The most

TABLE 3.11 Frequency and Geometric and Arithmetic Mean Levels of Selected Organochlorine Pesticide Residues in Human Adipose Tissue and Urinary Metabolites in the General Population of the United States, 1976

Type of Pesticide/Chemical	Percent Positive	Estimated Mean
Organochlorine in adipose tissue (ppm)[a]		Geometric
Total DDT equivalent	100	4.84
Hexachlorobenzene	93.4	0.06
Alpha BHC	0.3	—
Beta BHC	96.9	0.27
Gamma BHC (lindane)	0.4	—
Dieldrin	93.7	0.15
Heptachlor Epoxide	94.6	0.10
Oxychlordane	96.9	0.14
Trans-Nonachlor	93.4	0.10
Mirex®	0.3	0.18
Alkyl phosphate in urine (ppb)[b]		Arithmetic
Dimethyl Phosphate (DMP)	11.5	<20
Diethyl Phosphate (DEP)	7.9	<20
Dimethyl Phosphothionate (DMTP)	6.5	<20
Diethyl Phosphothionate (DETP)	10.8	<20
Diethyl Phosphodithionate (DEDTP)	0.2	<20
Phenolic in urine (ppb)[b]		Arithmetic
Pentachlorophenol	84.8	6.3
3,5,6-TC-1-P	16.1	< 5.0
2,4,5-TCP	1.7	< 5.0
p-Nitrophenol	1.7	<10.0
Silvex	0.2	< 5.0
Alpha Naphthol	13.8	<10.0
Isopropoxyphenol	1.3	<40.0
Carbofuranphenol	0.7	<40.0
3-Ketocarbofuran	1.0	<30.0
Other Metabolites in urine (ppb) (Malathion)[b]		Arithmetic
Alpha-Monocarboxylic acid	9.4	<30.0
Dicarboxylic acid	2.8	<30.0

[a]Kutz (1978).
[b]Kutz et al. (1978).

toxic pesticides are used in agriculture; hence, data on poisoning are heavily weighted with statistics from that sector.

The number of deaths from pesticide poisoning has declined significantly in the past 20 years. In 1974, however, there were still 52 accidental deaths (Hayes and Vaughn 1977). The number of suicides from deliberate ingestion or exposure to pesticides is about three times greater than the number of accidental deaths (Hayes and Vaughn 1977), and the total estimated mortality from pesticides annually is about 200 (U.S. EPA 1976). EPA estimates that 2,831 individuals suffering from pesticide poisoning are hospitalized each year (U.S. EPA 1976). In addition, there are approximately 12,220 emergency-room cases of pesticide poisoning each year (U.S. Consumer Product Safety Commission 1976). The number of additional cases treated on an outpatient basis by private physicians is conjectural.

The most reliable estimate of the number of cases of pesticide poisoning due to urban pest management practices comes from EPA's Pesticide Incident Monitoring System (PIMS). This nationwide, voluntary-reporting unit was established to document human and environmental pesticide poisoning incidents in the United States; data have been accumulated since 1966. At present, however, the exposure circumstances of any given incident and its possible health-related effects have not been verified by laboratory studies and epidemiological field visits. Only exposure incidents reported to PIMS through EPA regional offices and other centers are included, and therefore, the data constitute a lower bound on the probable number of cases. Between 1966 and October 1979, PIMS (unpublished data) had recorded 30,708 incidents, including poisoning by pesticides used in the home and garden.

PIMS provides useful preliminary data on many of the pesticides used in urban settings. Tables 3.12, 3.13, and 3.14 list the number of episodes associated with some of the more commonly used chemicals, and Table 3.15 provides additional detail on warfarin and naphthalene exposures. Most of the incidents resulted from ingestion, and 1,164 warfarin cases involved children. EPA has also recorded over 5,000 incidents in this approximately 12-year period involving chemicals reported to have occurred in the home (see Tables 3.16 through 3.18). These data illustrate the public health problem posed by pesticides, and the need for continued vigilance of new chemicals entering the market. In addition, the data in Table 3.18 also reflect the substantial misuse of pesticides in the home.

The effects of pesticides on human health are illustrated by the recent history of rodenticide use. Warfarin, the first anticoagulant rodenticide, was introduced commercially in 1950. It was followed by additional dicoumarol compounds and then a series of indandione compounds. For

TABLE 3.12 Number of People Affected by Pesticide Incidents Involving Dichlorvos (DDVP), Naled (Dibrom®), Trichlorfon, Tetrachlorvinphos (Gardona®), Malathion, and Diazinon at Home and Garden Sites, 1966 to June 1979

Pesticide Involved	Garden Site	Home Site	Total
Dichlorvos (DDVP)	0	239	239
Naled (Dibrom®)	0	14	14
Trichlorfon	0	6	6
Tetrachlorvinphos (Gardona®)	0	8	8
Malathion	2	256	258
Diazinon	0	418	418
Total:	2	941	943

SOURCE: U.S. EPA Pesticide Incident Monitoring System (PIMS) (Unpublished).

TABLE 3.13 Number of People Affected by Pesticide Incidents Involving Warfarin and Sodium Salt of Warfarin, by Site, 1966 to December 1978

Site	Warfarin Alone[a]	Warfarin in Combination[a]	Total
Home	137 + 1,002 = 1,139	1 + 37 = 38	1,177
Agricultural	0 1 = 1	0 + 1 = 1	2
Commercial	0 + 3 = 3	0 + 0 = 0	3
Recreational area	1 + 2 = 3	0 + 0 = 0	3
Industry	0 + 0 = 0	0 + 1 = 1	1
Unspecified	1 + 15 = 16	0 + 3 = 3	19
Public building	0 + 3 = 3	0 + 1 = 1	4
Total:	139 + 1,026 = 1,165	1 + 43 = 44	1,209

[a]Underlined figures represent sodium salt of warfarin incidents.

SOURCE: U.S. EPA Pesticide Incident Monitoring System (PIMS) (Unpublished).

nearly three decades rodent control in the United States has been based on these toxicants. In 1971, however, the discovery that rats (and later mice) had become resistant to these compounds stimulated a search for new rodenticidal compounds (Jackson et al. 1971, Jackson and Ashton 1979).

The only new rodenticide that has been successfully registered and

TABLE 3.14   Number of People Affected by Pesticide Incidents Involving
Naphthalene, by Site, 1966 to February 1979

Site	Naphthalene Alone	Naphthalene in Combination	Total
Home	163	22	185
Unspecified	3	1	4
Total:	166	23	189

SOURCE: U.S. EPA Pesticide Incident Monitoring System (PIMS) (Unpublished).

marketed between 1971 and 1979 is Vacor®, an acute (single-dose)
rodenticide (N-3-pyridylmethyl-N'-p-nitrophenyl urea) introduced in 1975
as a 2 percent bait formulation by the Rohm and Haas Company. Some
concern has also been evidenced about the toxicity of Vacor® to nontarget
animals. Cats are especially susceptible to Vacor®, and eventually
evidence of its danger to human beings became apparent from the use of
the baits in suicide attempts and from accidental ingestion.

It should be noted that poisoning from rodenticides constitutes only a
small fraction of all poisonings in the United States. Data from the
National Poison Center Network (1978) indicate that less than 1 percent
of the reported human exposure to toxic substances involves rodenticides
(see Note, Table 3.19), and Vacor® accounted for only 5 percent of these.
Children were less frequently involved with Vacor® than with other
rodenticides (see Table 3.19).

Concern about human ingestion of Vacor® relates in part to the fact
that this rodenticide causes irreversible destruction of the beta cells of the
pancreas (with subsequent development of diabetes mellitus) and auto-
nomic nervous system dysfunction (Peoples and Maddy 1979). From data
summarized from clinical reports prepared by Rohm and Haas between
1977 and 1979 (see Table 3.20), it became evident that serious and
permanent disability occurs frequently in survivors who used Vacor® in
unsuccessful suicide attempts or ingested it accidentally. This may be due
in part to the delay between ingestion and medical intervention. In 20
intentional ingestions, 5 deaths occurred 25 to 72 days after exposure.
Diabetes was not recorded in 28 cases involving children, probably because
ingestion by them is usually followed quickly by countermeasures. Only 1
suicide attempt in 17 was immediately successful, and only 1 case of

TABLE 3.15 Cause and Number of Home Pesticide Incidents Involving Warfarin and Sodium Salt of Warfarin[a] and Naphthalene, 1966 to December 1978, Including Human and Environmental Cases

Cause	Warfarin Alone and In Combination[a]	Total	Naphthalene Alone and In Combination	Total
Ingestion of pesticide	66[1] + 264; 0 + 11	341	71 + 15	86
Unspecified	53 + 730; 1 + 21	805	82 + 6	88
Suspected ingestion of pesticide			10 –	10
Ingestion of bait rodent control	5 + 12; 0 + 2	19		
Fume inhalation			1	1
Improper storage	2 + 1; 0 + 1	4		
Improper disposal	0 + 2; 0 + 0	2		
Dermal exposure	3 + 3; 0 + 0	6		
Suicide attempt	4 + 9; 0 + 3	16		
Ingestion of treated food	1 + 2; 0 + 0	3		
Suspect intentional poisoning	0 + 1; 0 + 0	1		
Multiple exposure	1 + 3; 0 + 0	4		
Improper placement	5 + 2; 0 + 0	7		
Ingestion of treated material	0 + 2; 0 + 1	3		
Bait spill	0 + 1; 0 + 0	1		
Improper application procedure			0 2	2
Total:	140 + 1032; 1 + 39	1212	164 23	187

[a]Underlined figures represent sodium salt of warfarin incidents.

SOURCE: U.S. EPA Pesticide Incident Monitoring System (PIMS) (Unpublished).

ingestion by a child was fatal—this because the parents were not aware of the circumstances and did not seek medical assistance.

Meanwhile, Peoples and Maddy (1978) reported 9 suicide attempts through ingestion of Vacor® in California (with 7 survivors developing diabetes mellitus and hypotension) and 12 accidental exposures of children. None of the children showed any symptoms of poisoning following treatment.

136

TABLE 3.16   Home Incidents Involving Accidental Exposure to Chlorfen-vinphos (Dermaton®)[a], 1966 to July 23, 1978

Age of Person	Number Affected	Exposure Route
12 mos.	1	Oral
18 mos.	1	Dermal, oral
5 yrs.	2	Oral[b]
		Oral
9 yrs.	1	Oral
Adult	2	Oral
		Dermal

[a]Dermaton® is used as a flea and tick dip for dogs.
[b]Product was mistaken for cough medicine in one case.

SOURCE: U.S. EPA Pesticide Incident Monitoring System (PIMS) (Unpublished).

Recent concern and comments about Vacor® 4 poisonings in the Chicago area prompted contact with the Cook County Hospital. Of four adult patients who had received treatment for Vacor® poisoning, three have died (Dr. W.D. Towne, Cook County Hospital, personal communication, 1979).

In May 1979 the production and sale of Vacor® was suspended by the manufacturer at the request of EPA because of the large number of accidental and suicidal poisonings resulting from its use in Korea and the United States, and the inability of the manufacturer to develop a container that was "both childproof and functionally satisfactory" (Rarig 1979).

Several new, "second generation" anticoagulant rodenticides and new formulations of older acute material have or will shortly appear on the market (Ashton and Jackson 1979, Jackson et al. 1978, Jackson 1979, and Kaukeinen 1979), but only some of them will be available to the homeowner.

There are also other ways in which pesticides can cause health and environmental hazards. When chlordane, for example, is used at very high dosages (up to 0.1 lb/10ft^2) as a soil poison for control of termites, persistent vapors from the treatment may enter forced-air ventilation systems and render premises uninhabitable; vapors have been observed to have the same effect following overzealous use of pentachlorophenol for timber impregnation (R.L. Metcalf, University of Illinois, personal communication, 1979). Dichlorvos, a relatively volatile residual fumigant, is a strong acetylcholinesterase inhibitor (Heath 1961) and a very active

TABLE 3.17 Number of Human and Environmental Incidents Involving Chemicals with Selected Active Ingredients, 1966 to December 1978

Active Ingredient	Total No. of Incidents	Home-related Incidents		
		Alone	In Combination	Total
Acephate	17	1	0	1
Aldicarb	90	7	1	8
Alkyldimethyl benzyl ammonium chloride (BTCs 776, 2125, 2125M, and 824)	6	0	2	2
Amitrole	25	2	1	3
Ammonium sulfamate	4	1	0	1
Arsenic	68	24	1	25
Arsenic acid	13	1	0	1
Arsenic trioxide	72	44	3	47
Benzene hexachloride	28	7	2	9
Boric acid	321	304	6	310
Cacodylic acid and sodium cacodylate	46	0	6	6
Calcium arsenate	20	1	12	13
Calcium hypochlorite	31	23	2	25
Captan	198	25	54	79
Carbofuran	94	6	0	6
Carbophenothion	11	2	2	4
Chlorfenvinphos	7	1	5	6
Chlorophacinone	9	6	0	6
Chloropicrin	110	0	7	7
Copper acetoarsenite	6	3	0	3
Coumafuryl	12	11	0	11
Coumaphos	9	2	0	2
Creosote	133	7	3	10
Dalapon	22	1	1	2
DEET	45	24	14	38
Dicrotophos	22	3	0	3
Dimethoate	106	3	1	4
Dioxathion	16	2	1	3
Disulfoton	111	26	8	34
DSMA	7	1	1	2
Endosulfan	91	2	1	3
EPN	36	0	1	1
Eptam	25	1	0	1
2-Ethyl, 1-3 hexanediol	9	7	2	9
Fensulfothion	44	1	0	1
Fenthion	29	3	2	5
Fonofos	39	3	0	3
Kelthane®	78	5	41	46

TABLE 3.17 Continued

Zinc phosphide	18	3	0	3
Total:	5,142	2,401	435	2,836
Lead arsenate	54	31	1	32
Lindane	242	70	64	134
Methoprene	1	1	0	1
Methoxychlor	88	2	49	51
Monuron	12	1	0	1
MSMA	44	4	1	5
Naphthalene	191	164	23	187
Naphthylthiourea	2	2	0	2
Oxydemetonmethyl	67	2	22	24
Paraquat and P. dichloride	318	36	2	38
Pentachlorophenol	162	27	24	51
Phorate	147	5	1	6
Propanil	54	8	1	9
Propham	1	1	0	1
Ronnel	26	4	10	14
Sodium arsenate	224	211	0	211
Sodium arsenite	100	37	4	41
Sodium hypochlorite	42	27	10	37
Sodium pentachlorophenate	13	0	1	1
Trichlorfon	33	4	1	5
Vacor®	32	29	1	30
Warfarin (and Sodium salt)	1,261	1,172	40	1,212
Zinc phosphide	18	3	0	3

SOURCE: U.S. EPA Pesticide Incident Monitoring System (PIMS) (Unpublished).

environmental mutagen (Voogd et al. 1972, Aswood-Smith et al. 1972) whose long-term consequences for persons and animals are unknown, as are those of the widely used fungicides captan and folpet, both of which are also very active mutagens (Legator et al. 1969). The fungicides benomyl and thiabendazole are teratogenic (Shternberg and Torchinski 1972, Seiter 1976, and Makita et al. 1973).

Some pesticides available for domestic use have been found to cause cancer in one or more strains of laboratory animal. These include chlordane, dicofol, and possibly the herbicide trifluralin because of nitrosamine contamination. The alkylating fumigants methyl bromide, ethylene oxide, ethylene dibromide, and dibromo-chloropropane (DBCP)

TABLE 3.18 Cause and Number of Home-related Human and Environmental
Pesticide Incidents, 1966 to December 1978

Cause	Alone	In Combination	Total
Ingestion of pesticide	946	132	1,078
Ingestion of pesticide bait	55	3	58
Improper storage	28	20	48
Improper application	11	6	17
Improper disposal	11	6	17
Spray application	8	14	22
Hand application	4	0	4
Hand spray application	15	28	43
Commercial application	2	3	5
Pesticide misuse	2	5	7
Pesticide application	12	17	29
Drift	8	7	15
Improper/inadequate/faulty protective gear or procedure	5	7	12
Re-entry	--	3	3
Contact with treated material	25	8	33
Spill/Splash	16	12	28
Topical contact	20	39	59
Mixing pesticide	0	2	2
Equipment failure/maintenance	4	5	9
Container failure/damage	6	2	8
Failure to follow label directions	8	3	11
Homicide/Suicide	46	12	58
Interior wood application	3	3	6
Exterior wood application	1	1	2
Wood treatment application	3	3	6
Disaster (fire, tornado, etc.)	2	2	4
Drainage/runoff	0	1	1
Unspecified	1,147	81	1,228
Pesticide unrelated	3	2	5
Total:	2,391	427	2,818

SOURCE: U.S. EPA Pesticide Incident Monitoring System (PIMS) (Unpublished).

TABLE 3.19 Description and Number of Incidents Involving Human
Exposure to Vacor® and Other Rodenticides in 1978

Circumstances	Exposures to Rodenticide Vacor®		Exposures to All Other Rodenticides	
	No.	Percent	No.	Percent
Age				
<10 years	23	57.5	649	87.1
>10 years	13	32.5	51	6.8
Unknown	4	10.0	45	6.0
Total:	40		745	
Container access				
Opened by child	3	13.0	149	23.3
Contents transferred	4	17.4	35	5.5
Container found open	0	–	72	11.3
Not in container	6	26.1	182	28.4
Unknown	10	43.5	202	31.6
Total:	23		640	
Location				
Own home	18	45.0	550	73.8
Relative/neighbor	2	5.0	42	5.6
Outdoors	3	7.5	14	1.9
Other	0	–	21	2.8
Unknown	17	42.5	118	15.8
Total:	40		745	
Severity of symptoms				
No symptoms	25	62.5	666	89.4
Mild - moderate	7	17.5	54	7.2
Serious - severe	5	12.5	10	1.3
Fatal	0	–	0	–
Unknown	3	7.5	15	2.0
Total:	40		745	

NOTE. Total reported human exposures to toxic or potentially toxic substances =
125,245. Rodenticide exposure accounted for 0.63% of these.

SOURCE: National Poison Center Network (1978).

TABLE 3.20   Rohm and Haas Company Data on Accidental Ingestion and
Suicide Attempts in United States with Vacor®, April 5, 1977 to April 4,
1979

		Symptoms or Outcome		
No. of Cases	No Major Symptoms	Diabetes Mellitus	Death	No Information
31 accidental ingestions (28 children and 3 adults)	23	1[a]	1	6[b]
17 attempted suicides (teenagers and adults)	1	10	2	4

[a]Adult.
[b]Age unknown in 2 cases; assumed to be children.

SOURCE: Chappelka (1979).

are not only suspect carcinogens, but DBCP can cause sterility in both
laboratory animals and human males (Whorton et al. 1977).

The spraying of residual pesticides—especially the organophosphate and
carbamate compounds—can also cause human poisoning inside homes.
Chlorpyrifos and diazinon are of marginal safety for use in households, but
they are used as cockroaches become resistant to materials less hazardous
to human beings. Home storage of stock solutions and unused spray
materials of toxic compounds with endosulfan, pentachlorophenol, paraquat, cyclohexamide, and even carbaryl and malathion, presents special
hazards to children.

Conclusion

The data reviewed above demonstrate that sizeable and multiple exposures
to pesticides can result from current urban pest management strategies.
EPA's National Human Monitoring Program for Pesticides has revealed
that incidental exposure of the general U.S. population to pesticides is
subtle but widespread. Data from the program, particularly the urinary
metabolite data, show that organophosphate and carbamate exposures do
occur through the use of pesticides in urban areas. In addition, data from
the Pesticide Incident Monitoring System (PIMS) suggest that human
pesticide poisonings occur in urban areas from time to time. In addition,
public health experience has demonstrated that insect resistance can result

in the resurgence of pest-borne diseases if control relies solely on chemicals.

These observations support the continued need for human monitoring programs as well as an expanded PIMS. With our concerns about cancer and birth defects, which can result from subtle exposure to certain chemicals, chronic exposure to pesticides has become as significant as acute exposure. Thus, the safe use of pesticides in urban areas will require increased vigilance, and future urban pest management plans should include alternative approaches that do not rely on chemical pesticides alone.

ENVIRONMENTAL EFFECTS

Pesticide use in the urban environment and in households represents a significant portion of total pesticide use in the United States (see Table 3.2), both qualitatively—in terms of the variety of products used—and quantitatively—in terms of the proportion of total pesticide production. Table 3.21 presents a list of insecticides, fungicides, and herbicides used in homes, gardens, orchards, and on turf and ornamentals. The table indicates uses, toxicity to rats and wildlife, and persistence in the environment.

Detailed characterization of the environmental effects resulting from urban and suburban use of the wide array of available pesticides is beyond the scope of this study. The environmental risk from any pesticide is a function of the pesticide's toxicity, persistence, biochemical action, and carcinogenic, mutagenic, teratogenic, or other hazardous characteristics. Moreover, the risk depends on the quantities used, the type of application, and the expertise of the user. As noted earlier, detailed information about these factors is difficult or impossible to obtain. Most of the pesticides used domestically are presumed to be in the general-use (nonrestricted) category. Certified pest control operators and public pest management specialists may employ highly toxic materials, however, such as the rodenticide sodium fluoroacetate, the herbicide paraquat, the fungicides cyclohexamide and pentachlorophenol, and the fumigants methyl bromide and sulfuryl fluoride.

Heavy use of herbicides in urban settings may cause damage to vegetable gardens and ornamental plants because of drift during applica- tion, or by runoff and leaching. Injudicious use of highly persistent herbicides can also inhibit the growth of domestic plants, sometimes for years.

Widespread use of soil insecticides to control moths and white grubs on lawns often has deleterious effects on earthworm populations and consequent unfavorable effects on soil structure. Soil contamination with a persistent insecticide like DDT following treatments for Dutch elm disease resulted in uptake of the chemical by earthworms and the subsequent decrease of local populations of robins and other birds (Barker 1958). Pets are frequent victims of unwise domestic use of pesticides. For example, dogs that feed on unprotected or improperly placed anticoagulant bait may be killed (Gates 1957; W.B. Jackson, Committee on Urban Pest Management, personal communication, 1980).

Perhaps the most insidious urban environmental effects of pesticides are leaching and runoff into the aquatic environment. The synthetic pyrethroids and endosulfan, chlordane, and methoxychlor are toxic to fish at ppm levels (see Table 3.21). These materials, along with pentachlorophenol and trifluralin, are lipophilic and consequently bioaccumulative through direct absorption from water or through aquatic food chains. Some herbicides, such as simazine, are highly toxic to algae and may affect the community structures of phytoplankton (U.S. EPA 1974).

ALTERNATIVE MANAGEMENT METHODS

The principles and practices of agricultural pest management are not entirely applicable to the management of many urban pests, particularly those affecting public health. For public health pests, the controlling principle must be to protect and maintain human health and life.

The management of pests that affect people should be controlled and guided by those whose primary responsibility is for public health and welfare, working in collaboration with those responsible for housing and urban development, for employment and labor, transportation, immigration, and other areas. There may be times, however, when it is also appropriate to consult those who are concerned with agricultural pest management.

In major emergencies the management of public health pests may require extreme measures: quarantines, mass treatment or vaccinations, or temporary destruction or suppression of pests, sometimes with the aid of measures whose damage to the environment must be weighed against the immediate need to protect human lives. Under more normal circumstances, the control of urban pests may involve direct treatment of affected people, or measures to improve personal habits and cleanliness. Control may also involve providing new or repaired dwellings, requiring landlords and tenants to follow certain sanitary or waste-disposal procedures, requiring employers to provide protection for employees, requiring new procedures in food-handling establishments, limiting the rights of builders

TABLE 3.21 Health and Environmental Effects of Pesticides in Common Use in the Urban Environment

Common Name and Use Code[1]	Rat Oral LD50 (mg/kg)[2]	Toxicity Rating[3]	Nontarget Hazard[4] Bird LD50 (mg/kg)	Fish LC50 (ppm)	Toxicity Rating	Environmental Persistence[5] Months	Rating	Cautions
INSECTICIDES								
Acephate G	866-945	2				>0.5	1	
Allethrin H,V	680-1,000	2		0.020+b	3	>0.5	1	
Bacillus thuringiensis G	>10,000	1		>10b	1	4-6	1	
Carbaryl G,T,V	307-986	2	>2,000p	2.0b	1	36-60	3	carcinogen
Chlordane S	283-590	2	1200 m	0.022b	2	3-6	5	
Chlorpyrifos H,T	97-276	3	8.4-17.7p	0.020b	4	3-12	3	
Diazinon H,G,T	66-600	3	4.3p	0.380b	2	>1	3	
Dichlorvos H,V	25-170	4			1		1	strong mutagen
Dicofol G,O	575-1,331	2		100b	1	24-60	4	weak carcinogen
Dimethoate G,O	250-500	3	41.7m	19+b	3	>2	1	
Endosulfan G	18-110	4	33m	0.003b	4		3	
Malathion H,G,V	885-2,800	2	1485m	0.130b	2	0.5	1	
Methiocarb G	130-135	3					1	
Methoxychlor G,O	5,000-7,000	1	>2000m	0.007t	3	6-12	3	estrogen
Naled A,V	430	2	52.2m	0.240b	2		1	
Oil spray O	>10,000	1			2		2	
Phosmet	147-299	3				2-6	3	
Propoxur H	95-104	3				1	2	
Propylthionopyrophosphate G	1,224-2,730	2				1	2	
Pyrethrins H,V	200-2,600	2	>10,000m	0.0545b	2	>0.5	1	
Resmethrin H,V	1500	2			2	>0.5	1	
Ronnel H	906-3,025	2				>2	2	
Stirofos V	4,000-5,000	1						
Trichlorfon T	450-469	2		32t	1	>1	1	

FUNGICIDES

Compound				I	II		III	Remarks
Benomyl G,O	>10,000			1				teratogen
Bordeaux O				1				mutagen
Captafol G,O	6,700			1		2-4	2	strong mutagen
Captan G,O	9,000			1				
Carboxin G	3,280			1				
Chloroneb G,O	>10,000	>2000m	0.150b	1				
Copper sulfate A,G,O	2.5	50-100m		5				no antidote
Cyclohexamide T	>10,000			1				
Dicloran O	980			2				
Dinocap G,O	1,000			1				
Dodine O	>17,000			1				
Ferbam G,O	>10,000			1				
Folpet G,O		>2,000m		1	1			
Lime-sulfur O	>8,000			1	1			
Maneb G,O,T	6,750			1	1			
Pentachlorophenol S	27-80	4,000-5,000p	0.1b	4	3	ca. 100	5	strong mutagen; oxidative phosphorylation inhibitor
Sulfur G,O	3,300			1	1			
Thiabendazole	780			1				
Thiram G,O	5,200	673p		2	1	2-4	2	
Zineb G,O	1,400	>2,000m,p		1	1	3	2	
Ziram G,O				1	1	1-2	2	

Table 3.21 Continued

Common Name and Use Code[1]	Rat Oral LD50 (mg/kg)[2]	Toxicity Rating[3]	Nontarget Hazard[4] Bird LD50 (mg/kg)	Fish LC50 (ppm)	Toxicity Rating	Environmental Persistence[5] Months	Rating	Cautions
HERBICIDES								
Amitrole A	14,700-24,600	1				0.5-1	2	
Chlorpropham	3,800	1	2,000m	10b	1	1-3	2	
2, 4-D A,T	300-1,000	2	1,000-2,000 m,p	250t	1	1-3	2	
		I						
DCPA O,T	3,000	1	2,000m	20t,b	1	2-3	2	
Dicamba G	2,900	1	673-800p	35b	1	3-12	3	
Dichlobenil A,O	3,160	1				2-6	3	
Diphenamid	830-1,100	2				3-6	3	
Diuron O	3,400	1	2,000m	4,300+b	1	3-12	3	
					II		III	
Oryzalin	10,000	1				3-12	3	
Oxadiazon	8,000	1				3-6	3	
Paraquat O	150	3		400b	1	0.5	1	dangerous, delayed toxin
Simazine A,O	5,000	1	2,000m	56+b	1	3-12	3	
Trifluralin	3,700	1		0.1+b	2	3-12	3	forms nitrosamines

NOTE. Use codes: A = aquatic   G = garden   H = household   O = orchard   S = structural   T = turf   V = veterinary
Fish and bird codes: b = bluegill   t = trout   m = mallard   p = pheasant

I. Mammalian Toxicity (rat oral LD_{50}, mg/kg)

1 = >1,000
2 = 200-100
3 = 50-200
4 = 10-50
5 = >10

II. Nontarget Hazard
(mallard or pheasant oral LD_{50}, mg/kg)

1 = >1,000
2 = 200-1,000
3 = 50-200
4 = 10-50
5 = >10

(blue gill or trout LC_{50}, ppm)

1 = >1.0
2 = 0.1-1.0
3 - 0.01-0.1
4 = 0.001-0.01
5 = >0.001

III. Environmental Persistence (average soil half-life)

1 = >1 month
2 = 1-4 months
3 = 4-12 months
4 = 1-3 years
5 = 3-10 years

SOURCE:

[1] Thirty-first Illinois Custom Spray Operators Training School (1979).
[2] Kenaga and End (1974); Spencer (1973).
[3] Metcalf (1975a).
[4] Pimentel (1971).
[5] Metcalf (1975a).

to develop new suburban areas, and requiring pet owners to control pets and their excreta.

Because of growing concern over sole reliance on chemical pesticides as a means of controlling pests, there is a need to examine more closely a variety of alternative pest management strategies. Although concern about pesticide use has revolved chiefly around agriculture, the Committee on Urban Pest Management feels that, for reasons explored elsewhere in this report, the need for alternative approaches may be even greater in urban areas. All of these alternative methods will require a sustained educational effort; some will require public financial assistance, while others will ultimately depend upon the police power of the community.

Following is a review of a number of selected alternative approaches, with particular emphasis on their potential applicability to urban pest problems. The reviews of three of these approaches—biological control, host-plant resistance, and genetic manipulation—are based on working papers provided to the Committee (see Appendix B).

BIOLOGICAL CONTROL

Biological control involves the use of natural enemies to maintain pest population density at a level lower than would occur otherwise. Natural enemies of pests include predators, parasites (or parasitoids), pathogens, and, in some cases, nonpest competitors. In agriculture, successful biological control is achieved when natural enemies maintain pest population density below the economic injury level. Successful biological control in urban areas, however, is achieved when pest population density (or some other suitable parameter) is kept below the level at which aesthetic or public health injury occurs.

Biological control, like other methods of pest control, is best considered as a particular tactic rather than a full-scale plan or strategy. Biological control is an ecologically sound technique, and when combined with other appropriate tactics the result is integrated pest management (IPM). In some cases, biological control by itself may provide permanent pest suppression.

The basic premise of biological control can be summarized as follows: In the native home or area of origin of a given pest, there exists a natural enemy or a complex of natural enemies that maintains (or is capable of maintaining) the particular pest at comparatively low levels. This phenomenon is termed "natural biological control." On the other hand, pests that exist in the absence of their adapted natural enemies and that eventually reach outbreak proportions are subjects for "classical biological control"—introduction of the appropriate natural enemy or enemies, usually from the native home. Both native and exotic (i.e., introduced) pests are amenable to classical biological control, although exotic species have traditionally received most of the public and scientific attention. An important component of biological control is conservation and augmentation of a pest's natural enemies. Conservation normally involves the use of selective insecticides or selective use of nonselective insecticides. Augmentation means manipulation of a pest's natural enemies and their environment in order to increase their effectiveness. Natural biological control, classical biological control, and conservation/augmentation techniques constitute one modern approach to biological control of arthropod pests.

Natural Biological Control

Many ecologists assume that urban flora and fauna show little diversity (see, e.g., Stearns 1978). Although this may be true of certain taxa (e.g., vertebrates), it is clearly not true for plants or for insects and other arthropods (see references in Frankie and Ehler 1978, Davis 1978). In fact, the variety of arthropod species in certain urban environments may be comparable to—if not greater than—that of nonurban environments (see, e.g., Ehler and Frankie 1979a, b; see references in Frankie and Ehler 1978). Since many urban insect pests may be attacked by natural predators and parasites, it behooves entomologists to determine the impact of these natural enemies.

Although thorough analysis of natural biological control in urban areas remains virtually nonexistent, one particular study does warrant consideration. This is the classic work of Luck and Dahlsten (1975), who examined the population ecology of a pine needle scale (*Chionaspis pinifoliae* Fitch) in South Lake Tahoe, California. The goal of their study was to find out why an infestation of the scale was generally confined to residential areas. The study revealed that the scale was generally under effective natural biological control in the surrounding forests, but that insecticide fogging to control mosquitoes in the city had also destroyed the natural enemies of the scale, thus causing it to increase. The scale declined to its natural level following cessation of the mosquito control program.

There is no reason to assume, a priori, that natural biological control in the urban environment cannot be as effective as it is elsewhere. With this in mind, it is worthwhile to consider the use of native (endemic) plants in urban landscapes. From an entomological point of view, native plants could be of great value when native plant-feeding insects associated with the plants are maintained below aesthetic injury levels by native natural enemies—i.e., through natural biological control.

This point is illustrated by coyote brush, *Baccharis pilularis* Decandolle, which is endemic to coastal and certain inland areas of California. Two distinct forms of *Baccharis* exist: a prostrate form, subspecies *pilularis* (= *typica*), that occurs only on a narrow coastal strip in northern California; and subspecies *consanquinea*, an erect shrub that occurs along the coast and at inland locations (Doutt 1961). The prostrate form is often used as a ground cover in urban landscapes and along the margins of California freeways.

Since it is a native species, *Baccharis* is well adapted to the Mediterranean climate that characterizes much of California. In urban areas the plant usually obtains sufficient water from winter rains to permit growth and maintenance; thus it is able to pass the dry summer without irrigation. This, of course, is advantageous from a horticultural point of view.

A cecidomyiid midge (*Rhopalomyia californica* Felt), which induces galls on the terminal branches of *Baccharis*, has been the subject of a long-term ecological investigation in the city of Davis, California (L.E. Ehler, University of California, personal communication, 1979). The midge is naturally controlled in nonurban settings by 10 species of parasitic Hymenoptera (Doutt 1961, Force 1974), only 6 of which are associated with the midge in Davis. These 6, however, are the same species that characteristically occur with the midge throughout the state, and the degree of biological control obtained thus far in Davis appears to be comparable to, and as aesthetically acceptable as, that observed in the natural habitat.

There are at least two ecological questions that must be answered, however, with respect to the use of native plants. First, can the arthropod communities associated with native plants in urban environments be expected to be structurally similar to their nonurban counterparts? This is especially crucial in the case of guilds of natural enemies. In this regard there is evidence that such structural similarity can be expected (Ehler and Frankie 1979a, b), although considerably more corroborative evidence is required. Second, given at least some degree of structural similarity in a particular guild of natural enemies, can comparable degrees of natural biological control be anticipated? Such biological control may be obtained rather quickly, as in Davis, or it may be obtained after a much longer period of time has elapsed (see, e.g., Frankie et al. 1977, Frankie and Ehler 1978).

Classical Biological Control

The urban environment abounds with exotic plant species, including numerous associated exotic pests that have invaded without their natural enemies. For this reason classical biological control is well suited to some urban pest problems.

Little attention has been devoted, however, to the introduction of natural enemies of insect pests in the urban environment in this country. This becomes apparent in analyzing the attempts at classical biological control of insect pests in the United States from about 1890 through 1968 (Clausen 1978).

In one recent attempt, 15 urban pests were the targets of classical biological control (Laing and Hamina 1976). (These do not include agricultural, medical/veterinary, forest, and greenhouse pest species that also occur in urban areas.) To combat the 15 pests, 70 species of parasites and predators were released, 24 (34.3 percent) of which became established. Substantial levels of control (none complete) were obtained in 3 cases: nigra scale (*Saissetia nigra* Nietner), lecanium scale (*Lecanium tilae* Linn.), and oriental moth (*Cnidocampa flavescens* Walker). Substantial success occurs when insecticide treatments for the pest in question are reduced considerably following establishment of a natural enemy; similarly, complete success results in virtual elimination of insecticide treatments for the pest in question.

In contrast, 106 agricultural, medical/veterinary, forest, and greenhouse pests were the targets of another experiment in classical biological control. To combat these pests, 915 species of predators and parasites were released, of which 269 (29.4 percent) became established. Numerous successes (some complete) were achieved, particularly in the agricultural sector (Huffaker and Messenger 1976).

During the late 1960s the Nantucket pine tip moth was accidentally introduced into Santee, San Diego County, California, in a shipment of Monterey pines originating in Tifton, Georgia. By 1978 the moth had spread to most areas in the county where Monterey pine was planted as an ornamental, and it also became established in Orange County. The Nantucket pine tip moth affects three commodities: choose-and-cut Christmas trees, nursery stock, and ornamental plantings.

In 1974 two natural enemies of the moth, *Campoplex frustranae* (Hymenoptera: Ichneumonidae) and *Lixophaga mediocres* (Diptera: Tachinidae) were imported from the southeastern United States. *Campoplex frustranae* became established and now has reduced the moth population to an extremely low density at the original site of infestation. This natural enemy of the moth is now spreading to other infested areas in San Diego and Orange Counties. Monthly samples taken before and after parasite establishment show the degree of control achieved on the sample trees in the initial release site. The results generally reflect the pattern of infestation in the area immediately surrounding the site where natural enemies were released. Approximately 3 to 4 years will be needed, however, before *Campoplex* achieves its potential effectiveness.

Since 1968 there have also been some efforts devoted to classical biological control of insect pests in urban areas (see, e.g., Olkowski et al. 1976, 1978b). However, there is much to be done, and it is strongly recommended that introduction of natural enemies of urban pests receive increased attention and financial support.

Biological control of weeds with insects has been described as the inverse of biological control of plant-feeding insects. That is, rather than introducing natural enemies of an accidentally imported plant pest (as in classical biological control), the objective in biological control of weeds is to intentionally introduce a host-specific plant-feeding insect (pest) without its natural enemies. In this way the imported weed-control insect is freed of biotic restraints that might otherwise prevent it from having a significant impact on the weed in question. More detail on biological control of weeds and the impact of weeds as urban pests is provided in the working paper, Urban Pest Management—Weed Science (see Elmore, in Appendix B).

Conservation and Augmentation

Conservation, as used here, means enhancing the efficacy of a pest's natural enemy through manipulation of the enemy itself or through modification of the environment. These procedures are applicable to both native and introduced species and have received considerable attention, particularly in agricultural pest management (DeBach and Hagen 1964, van den Bosch and Telford 1964, Rabb et al. 1976, Ridgway and Vinson 1977). Conservation generally involves harmonious use of chemical controls as well. Such approaches should be applicable to many urban pests. In addition, it should be noted that under some circumstances natural enemies themselves may also become pests.

Probably the most common method of manipulation of natural enemies is inundatory release of massive numbers of the enemy or enemies in order to effect immediate pest suppression. The goal is to use a pest's enemy or enemies as a kind of "biotic insecticide," although in some cases pest suppression may not occur as rapidly as with the use of an effective chemical insecticide. A number of natural enemies of pests are commercially available to urban residents for this purpose. They include *Chrysopa carnea* Stephens for control of aphids, several species of phytoseiid mites for control of spider mites, *Encarsia formosa* Gahan for control of greenhouse whiteflies, and *Hippodamia convergens* Guerin for control of aphids. The use of microbial insecticides also falls into this category. Two examples are *Bacillus thuringiensis* Berliner, effective against many species of lepidopterous larvae, and *Bacillus popilliae* Dutky, effective against certain scarab larvae.

The utility of these natural enemies, however, is variable. For example, use of *Bacillus thuringiensis* is an effective means of controlling California oakworm (*Phryganidia californica* Packard), and because it is safe and selective it fits well in an integrated pest management program (Olkowski et al. 1974, Pinnock 1976, Pinnock et al. 1978). Inundatory releases of *H. convergens*, on the other hand, are generally of limited value, apparently because the beetles are genetically programmed to disperse once released. Thus, there is clearly a need to assess the various natural enemies of pests that are commercially available to urban residents to determine which ones can be considered acceptable alternatives to chemical insecticides.

Environmental modification to enhance the efficiency of pests' natural enemies can also lead to increased biological control in urban areas. A number of techniques are currently available. These include "food sprays" designed to increase egg production in *Chrysopa carnea* Stephens (Hagen et al. 1971); sticky traps for controlling Argentine ant (*Iridomyrmex humilis* Mayr) and thus increasing the effectiveness of certain natural enemies of honeydew-producing Homoptera; and planting "insectary crops" to attract and maintain predators and parasites.

Efficacy data for many of the techniques of environmental modification are unavailable. Where efficacy tests have been conducted—e.g., the application of food sprays to an urban community garden (G.D. Propp and L.E. Ehler, University of California, personal communication, 1979) and the use of insectary plants in small garden plots (W.F. Crepps, University of California, personal communication, 1979)—the results have often been inconsistent and unclear. There is considerable need for additional research into the applicability of such measures in the urban environment.

Research Priorities

Despite its potential usefulness, biological control of urban pests has received relatively little attention, and use of the technique—particularly to control public health pests—should be investigated. Priority in research programs should be given to pests that range over many states, including such well-established introduced species as the elm leaf beetle (*Pyrrhalta luteola* Muller), the Japanese beetle (*Popillia japonica* Newman), and the American cockroach (*Periplaneta americana* Linn.). Native pests with widespread harmful effects include the fall webworm (*Hypantria cunea* Drury), the corn earworm (*Heliothis zea* Boddie), and the southern chinch bug (*Blissus insularis* Barber). Some priority should also be given to pest problems of regional significance—e.g., the exotic *Pulvinaria* spp. that infest vast acreages of ice plant in northern California, or the native California oakworm that infests live oak in the same region, or the cluster fly that hibernates in the northeastern United States.

Consideration might also be given to pests that are important in both the urban and nonurban areas. The gypsy moth (*Porthetria dispar* Linn.), for example, is a pest of both forests and urban shade trees. Similarly, black scale (*Saissetia oleae* Oliv.), a pest of citrus and olive trees, is also a pest of ornamental plants in urban environments. The housefly (*Musca domestica* Linn.) is an example of a pest that is of medical/veterinary importance in both urban and nonurban areas. Other pest species that are important in both urban and suburban areas include the spruce budworm (*Choristoneura fumiferana* Clemens), white grubs (*Phyllophaga* spp.), and the tomato hornworm (*Manduca uinquemaculata* Haworth).

Further documentation of the value of natural biological control is sorely needed, since such knowledge is of particular importance with respect to the development of integrated pest management programs, many of which are based on natural enemies of pests. Future research might also be directed toward the entomological aspects of using native plants in urban areas, particularly to identify plants that are suitable and those that are not.

Finally, priority must be given to critical evaluation of the effects of biological control and IPM programs in urban environments.

HOST-PLANT RESISTANCE

The selection of ornamental plants for use in urban landscapes historically has been based primarily on such aesthetic characteristics as color, shape, height, and shade-giving potential. But in more recent years greater emphasis has been given to such practical considerations as leaf- or fruit-shedding characteristics; root behavior in relation to curbs, sidewalks, and sewage systems; and conformation with respect to pruning and topping needs. The impact of such devastating diseases as Dutch elm disease, elm phloem necrosis, and oak wilt has also been a factor in selection. More recently, environmental and human health concerns about pesticides, especially insecticides, have raised the possibility that ornamental plants in urban areas might also be chosen on the basis of their resistance to pests.

The Scope of Resistance

Each ornamental plant species is host to only a relatively small proportion of the thousands of insects and mites considered to be pests. It follows, then, that each plant species is immune to the vast majority of pests. The chinch bug, for example, does not attack any plant species but those of the grass family; red pine scale does not feed on any tree species other than certain pines; and the California oakworm does not infest cypress, poplar, or turfgrass. Host-plant resistance, also termed nonhost immunity, is likely to be a permanent barrier to pest infestation.

Along the continuum from immunity to great susceptibility are plant species that exhibit varying degrees of resistance to pests. Such resistance may be based on one or a combination of mechanisms, including nonpreference, antibiosis, and tolerance (Painter 1951). Resistance, unlike immunity, is not likely to be permanent, although in most cases it is a goal well worth pursuing. It is important to note that in the so-called "breaking down" of resistance it is usually not the plant that undergoes change; instead, it is change on the part of the pest that permits exploitation of the host. Particularly informative papers on the subject of pest-resistant ornamentals have been written by Gerhold et al. (1966), with emphasis on forest trees; Weidhaas (1976), and Morgan et al. (1978).

Growers of landscape ornamentals have a broad choice of vegetation from which to select for a given site. There are literally dozens of varieties of turfgrass and other ground covers, and hundreds of species, varieties, and cultivars of shrubs and trees. Where the pests of a given ornamental are known or perceived to exceed a tolerable level, selection of a more pest-resistant plant (as opposed to development of an alternative through a plant-breeding effort) can reduce the need for maintenance or replacement (Felt 1905). The breeding of ornamentals for the purpose of developing specific pest-resistant vegetation has some precedent, however, particularly in turfgrass (Morgan et al. 1978).

Examples of Progress in Identifying Pest-resistant Ornamentals

Munro (1963) evaluated 40 species and cultivars of *Ceanothus* in an arboretum for susceptibility to the ceanothus stem-gall moth, *Periploca ceanothiella* Cosens, an insect whose galls reduce the flowering capacity of the plant. More than half of the plants showed no occurrence of the insect. Severe infestations were noted only on *Ceanothus griseus* and its horticultural selection *horizontalis*, with the remainder of the infested plants showing light to moderate occurrence.

Munro (1965) published the results of a similar evaluation of the resistance of 112 species of *Acacia* and 6 of *Albizia* to the acacia psyllid, *Psylla uncatoides* Ferris and Klyver. This insect causes chlorosis of the foliage and tip dieback on susceptible hosts. As in the case of *Ceanothus*, a majority of the plants showed little or no occurrence of the psyllid. Such *Acacia* species as *baileyana*, *dealbata*, *podalyriaefolia*, and *verticillata* are indeed immune to the pest, as confirmed by observations and critical evaluation.

Williams et al. (1977) screened eight species and varieties of *Euonymus* to determine their susceptibility to the euonymus scale, *Unaspis euonymi* Comstock, in Alabama. The scales failed to survive on *E. kiautschovicus*, whose use is recommended where low-maintenance plants are desired.

The cypress tip moth, *Argyresthia cupressella* Walsingham, causes unsightliness to many species of Cupressaceae along the Pacific coast. Sixteen species and cultivars of *Chamaecyparis*, *Juniperus*, and *Thuja* were evaluated over a 2-year period for resistance to the pest (Koehler et al. 1979); seven were found to exhibit low susceptibility and are now recommended for planting in coastal California.

Problems and Limitations in Developing Pest-resistant Ornamentals by Selection or Breeding

Weidhaas (1976) identified the principal problems encountered in developing pest-resistant ornamentals:

1. Often, relatively few specimens of specific ornamentals are grown in a given area, which exacerbates the problem of selecting truly resistant ones. Pseudoresistance, the apparent resistance in potentially susceptible plants, resulting from host evasion, induced resistance, or escape, is likely to be a problem when few plant specimens are available to express (or not express) the genetic variation of a species.

2. The great diversity of insect pests that attack ornamentals and the increase in the number of pests as the ornamental becomes more widely planted, with the result that pest problems may outstrip the capacity to deal with them by developing resistant ornamentals.

3. Fluctuations in local populations of major and minor pests, some of which are the result of changes in cultural and pest control practices.

4. Lack of research personnel and funds.

Pest-resistant ornamentals will, of course, solve none of the problems associated with already established plantings. It is only at new and replacement sites that they are likely to have any benefit. Yet the generally good economic health of the ornamentals industry in the leading producing states of California, Florida, Oregon, Ohio, and New York suggests that a great many new plants are being produced and sold.

The most useful ornamental plants will be resistant not only to insect and mite pests, but also to plant diseases as well as other agents responsible for poor appearance and performance. The question of how a pest-resistant ornamental plant can gain general acceptance demands attention. The research required to identify or otherwise develop resistant plants does no good unless a financially motivated private sector and an informed public do their part. The public must learn, particularly in the case of nonnative ornamentals, that the longevity of some kinds of resistance cannot be guaranteed. People have only limited control over the introduction of new pests into an area, and no control over the consequences that may arise from planting a "resistant" plant outside the region where resistance has been demonstrated (Weidhaas 1976). The retailers of ornamentals should be constrained from advertising "resistant" plants in a way which implies that resistance is tantamount to immunity from pests.

Conclusion

The problems in developing pest-resistant ornamentals were enumerated by Weidhaas (1976) at a time when chemical pest control was viewed somewhat more favorably than it is today. Public resistance to the use of pesticides in urban areas is growing, and greatly reduced use of sprayed pesticides may be anticipated.

Selection of ornamental plants is increasingly being made on utilitarian grounds, and pest resistance seems likely to be given greater emphasis in the decision-making process. Some pest-resistant plants are already known, and many more could be identified within the next few years.

INSECT GROWTH REGULATORS

Insect growth regulators (IGR) are chemical insecticides that interrupt the complex growth and metamorphosis patterns of insects between immature molts or during pupation. There are two general types of IGRs, juvenoids and chitin development inhibitors. The juvenoids are structurally optimized analogs of the insect growth hormone neotenin. The best-known juvenoid is methoprene, or isopropyl (E,E)-11-methoxy-3,7,11-trimethyl-2,4-dodecadienoate, which is particularly effective against mosquito larvae; it is active at ppb levels and prevents emergence of the adult from the pupa, thus curbing the obnoxious stage. Methoprene is also registered as a fed-through agent to prevent the development of horn flies (*Haematobia irritans*) in the excrement of cattle. Recent work also indicates that it has potential for controlling fleas (Chamberlain 1979).

The chitin development inhibitors are best represented by diflubenzuron, or N-(4-chlorophenyl-N'-2,6-difluorobenzoyl) urea, which interrupts the formation of chitin from acetylglucoseamine, thus interrupting egg development, molting, and pupation. Diflubenzuron at ppb levels is effective against mosquito larvae and against such other pests as the gypsy moth *Porthetria dispar* and the cotton boll weevil *Anthonomus grandis*.

Insect growth regulators are generally highly specific against insects because their biochemical lesions are processes unique to arthropods, i.e., molting and retention of juvenile characters, and chitin formation. IGRs also pose little risk to mammals. Rat oral LD_{50} of methoprene, for example, is greater than 34,600 mg/kg, and of diflubenzuron is greater than 4,640 mg/kg (Kenaga and End 1974). Methoprene is moderately persistent, and diflubenzuron is quite persistent in the environment. Both are effective only against immature stages of insect pests and are very slow in toxic action, characteristics that present major drawbacks for use against many urban insect pests. In addition they currently are expensive (i.e., at $50-60/lb) and thus they appear to have limited usefulness in urban pest management programs.

GENETIC MANIPULATION

Genetic manipulation is another method of controlling or managing urban pests, but for a significant number of pest species this approach has not yet been successful. Perhaps the best-known successful effort is the sterile-male release program for the screw-worm fly. Other attempts at genetic manipulation have involved the Mediterranean fruit fly, the olive fly, several species of mosquitoes, and the German cockroach. None, except for the screw-worm fly, has reached the point of widespread use. A recent Rockefeller Foundation publication (Hoy and McKelvey 1979) contains a number of papers that address the application of genetics to significant problems in management of insects and related arthropods.

There are several approaches to genetic manipulation, but sterilizing male insects by exposing them to ionizing radiation or to chemosterilants is most often tried. If the species can be mass-produced, large numbers of sterilized males can be released into the field, thus reducing pest reproduction there.

Male sterilization is not simple, however, and there are many complicating factors. Successful mass production depends on a suitable laboratory diet, which may be difficult to develop. Subtle genetic changes may occur in a laboratory colony rendering sterile males unacceptable to females in the field population (Bush et al. 1976). Experiments with the German cockroach have shown that the dose of radiation necessary to produce sterility also produces considerable short-term mortality (Ross and Cochran 1963). This may mean that the surviving insects have been too damaged to be good competitors in a field population.

Other types of genetic manipulation include reciprocal chromosome translocations, chromosomal inversions, and cytoplasmic incompatibility. All three have been suggested for possible use, particularly with mosquitoes. Of these alternatives, chromosome translocations have received the most study (Cochran 1976, 1977; Cochran and Ross 1977a; Ross and Cochran 1975a).

Utilization of chromosome translocations in the German cockroach required a formal genetic and cytogenetic base of information which took 15 to 20 years to develop (Cochran and Ross 1969, 1974, 1977b; Ross and Cochran 1966, 1971, 1973, 1975b, 1976, 1977, 1979). Development of similar knowledge about other species, depending on their genetic make-up, could take more or less time. The induction of genetic markers is slow but not necessarily difficult. The meiotic chromosomes of the German cockroach, however, are very amenable to study (Cochran 1976, 1977).

There are also many other biological factors that must be considered prior to undertaking a genetic control program for an urban pest. In an inundatory release program, for example, it is useful if females mate only once (Cochran 1979). On the other hand, a program to raise masses of pests may be difficult if the species produces only one generation per year. Thus, a thorough knowledge of species biology should be available before a decision is made on which species to select for control by genetic means.

Another problem is whether or not opportunities for releasing genetically altered insects into the urban environment exist or can be developed. Human attitudes on this question are significant. The sterile male screwworm flies mentioned earlier were released primarily in rural areas by aircraft, and people rarely saw them. Releasing sterile insects into homes is another matter altogether, although under certain circumstances it might be possible to convince people to allow it.

Other questions also arise regarding genetic manipulation programs. Would people interfere with the program by using insecticides when requested not to? Are they willing to provide an accurate picture of the pest problem in their homes? Would they be willing to accept small-scale residual infestations in their homes, since genetic manipulation might not totally eradicate a pest? Problems like these might be dealt with by including an appropriate educational effort in the release program.

Despite the impediments discussed above, genetic control of urban insect pests is still a valid prospect, provided that the selection of species to be used is made intelligently, and that both the public and elected officials understand that it could take a long time and be very expensive. What should be emphasized is that genetic manipulation might prove to be a much safer alternative to continued heavy use of chemical pesticides.

HABITAT MODIFICATION

The density of a given pest population can be increased or decreased by manipulating some of the limiting factors of its environment such as food, shelter, and predation. Reduction of a pest population by predation alone, or by the use of pesticides, may only be temporary, however. An alternate approach is habitat modification, that is, the elimination of conditions conducive to pest infestations. Habitat modification supplemented with other strategies can achieve a lasting degree of pest management.

In urban environments habitat modification may include such activities as:

1. Proper storage, collection, and disposal of organic wastes and other refuse;
2. Good housekeeping in homes and in institutional, industrial, and business establishments;
3. Proper storage and handling of food;
4. Elimination of harborage for pests indoors and outdoors through cleanup and proper disposal of debris and refuse from yards, cellars, vacant lots, and other such places; proper design, construction, and installation of food service equipment; and proper building construction and maintenance;
5. Demolition of abandoned and dilapidated structures;
6. Enforcement of sanitation, health, and housing codes; and
7. Permanent elimination of pest breeding sites and other sources of pest infestation.

These activities, also referred to as environmental sanitation, are most applicable in the management of a number of common urban pests, such as rodents, cockroaches, flies, termites, food and fabric pests, and domestic mosquitoes.

(Habitat modification for weed control is discussed in the working paper, Urban Pest Management—Weed Science (see Elmore, in Appendix B).)

The primary goal of habitat modification—the removal of aspects of the environment that are conducive to the propagation of pests—has special appeal and relevance in urban environments, particularly in dilapidated, filthy inner-city areas. Chapter 1 of this report alludes to some of the factors and symptoms of decay that affect pest populations in inner-city neighborhoods. Such conditions are clearly responsible in part for the occurrence and prevalence of a number of rodent and arthropod species that affect the health, quality of life, and general well-being of millions of poor people in American cities. Health statistics demonstrate causal relationships between morbidity and mortality and physical surroundings, and bear out the contention that people living in economically depressed neighborhoods are less healthy than those living in other areas (U.S. DHEW 1977).

Habitat modification needs more attention than it has received to date. Although likely to have its greatest impact on inner-city areas, habitat modification may also be relevant to pest problems in suburban homes (Frankie and Levenson 1978, Frankie and Ehler 1978), urban and suburban recreational areas, and food-handling establishments (NPCA 1979a). Research should focus on ecological relationships between pests and their habitats, pest-borne diseases, improved sanitation, and human socio-behavioral characteristics.

CULTURAL, MECHANICAL, AND PHYSICAL CONTROL OF ARTHROPODS ON PLANTS

Numerous nonchemical methods have been used to reduce herbivorous arthropod populations. Although many of the methods are applicable to both indoor and outdoor plants, the emphasis has been on outdoor plants. The methods can be conveniently sorted into two groups: (1) cultural and (2) mechanical and physical (Lawless and von Rumker 1976).

Cultural Methods

A variety of cultivation methods and other approaches aimed at reducing pest populations are included in this category. Common methods include (1) proper selection of plants for particular geographic areas and for particular habitats within given urban areas; (2) proper preparation of soils; (3) establishment of appropriate water and fertilizer schedules, and proper selection of fertilizers; (4) use of companion and trap plants; (5) early planting dates; and (6) annual rotation of plants. Insight into lesser-known or esoteric cultural controls can be gained by examining anecdotal accounts of ecological interactions between pests and host plants in Johnson and Lyon (1976).

With few exceptions (e.g., Latheef and Irwin 1979, Tahranainen and

Root 1972), cultural controls have received little attention from scientific investigators. These methods have received widespread publicity in organic gardening publications, however, and they are beginning to receive considerable attention in extension service pamphlets and popular garden publications (e.g., the *Sunset* magazine that is circulated widely in the western United States). The frequency and extent to which cultural methods of pest control on plants are used by urbanites are unknown. However, a questionnaire sent to 700 urbanites in California, Texas, and New Jersey indicated that nonchemical methods have been tried by many people, many of whom were generally satisfied with the results (Levenson and Frankie, in press).

Mechanical and Physical Methods

Common mechanical and physical methods that have been used by urbanites against certain pest groups are listed below.

Method	Pest
1. Washing or flooding with water	Washing: aphids, thrips Flooding: grubworms (scarabs)
2. Traps of various types	Sticky cards for whiteflies Rolled paper for earwigs Light traps for moths
3. Barriers and sticky bands	Certain caterpillars
4. Hand-picking	Bagworms, some other caterpillars, and snails
5. Pruning	Tent caterpillars, fall webworms, gall-inducing arthropods
6. Destruction of plant materials with persistent pest problems	Numerous types

As with cultural methods, little technical information exists on mechanical and physical methods. These have received attention from the popular press, however, and many urbanites are believed to have tried at least some of them. This is understandable, since a number of the methods rely on intuition (Levenson and Frankie, in press). Future research should attempt to develop more precise information on the application and limitations of these methods.

APPLICATION OF INTEGRATED PEST MANAGEMENT TO URBAN PEST PROBLEMS

Integrated pest management (IPM) is an attempt to maximize the forces of natural pest control and to use other tactics, such as pesticides, to the smallest degree possible (Glass 1975). More specific definitions and a history of IPM are found in Volume 3 of NRC's *Principles of Plant and Animal Pest Control* (1969). Although it originally evolved in agriculture, the basic philosophy of IPM also is applicable to some urban pest problems.

Integrated pest management is based on principles of applied ecology, and the ecosystem is generally used to implement a given IPM scheme. This requires an understanding of the interacting relationships (including human activities) within an ecosystem. Various segments of the urban environment (such as buildings and city parks) can be identified as ecosystems (Stearns and Montag 1974, Frankie and Koehler 1978). Furthermore, the concept of urban ecosystems has been successfully used for managing certain urban arthropod pests (Olkowski et al. 1976, Piper and Frankie 1978).

In agricultural pest management there is a threshold above which pest populations will cause serious economic damage to crops. In urban pest management there may be an economic, aesthetic, or public health threshold (NRC 1969) above which the effects of pests will not be tolerated. Despite the often low or zero injury level in urban pest management, it is still possible in many instances to realize an integrated effort.

Integrated programs usually evolve slowly (NRC 1969) because pest situations and ecosystems are complex, and it takes time to identify and characterize relevant interactions before an integrated effort can be implemented. In agriculture, a systems analysis approach is being used successfully to analyze and deal with complex pest problems, and it is expected that in the future this methodology will be applied to urban areas as well. However, much needs to be learned about interaction between pest organisms and human attitudes and ecology (including the influence of socioeconomic and political variables) before urban pest situations can be managed in an integrated fashion.

There are many situations in the urban environment where the integrated approach can be implemented, but it is also clear that in some cases a truly integrated effort cannot be realized. A unilateral rather than an integrated approach must often be used, for example, where a pest must be eradicated.

Selected examples of integrated pest management programs are discussed below.

RODENT CONTROL IN URBAN AREAS

Rats and mice are among the most successful mammal pests in urban areas. The Norway or brown rat (*Rattus norvegicus*), the roof rat (*Rattus rattus*), and the house mouse (*Mus musculus*) originated in Asia and later spread to all parts of the world. Both the Norway rat and the house mouse are widely distributed throughout the United States, while the roof rat is mostly found in coastal and southern areas of the country.

Reliable estimates of the domestic rodent population are not available. The frequently cited figure of one rat per person, originally derived from data about the number of rats on farms in England, is without foundation (Jackson 1977). A rodent population's size, which is the result of interaction between population forces and limiting environmental factors, such as food, harborage, and predation, will increase or decrease following changes in these elements. In 1947, for example, the rat population of Baltimore was estimated to be about 165,000 (Davis and Fales 1950). Two years later, following improvements in housing and in refuse collection, the estimated number declined to about 60,000, or approximately one rat per 15 inhabitants (Davis and Fales 1949). In New York City, ratios of up to one rat per 36 people have been reported (Davis 1950). These data constitute the only estimates derived from actual censuses of rat populations in American cities.

Although rats are associated with many pathogens and parasites (see Chapter 2), rat bites are the greatest threat. Rat bites, most frequently of children, are more common in inner-city neighborhoods than anywhere else. Before 1964, when the present rodent control program was initiated in New York City, the average number of rat bites reported to the city's health department each year was 693 (Raphael 1972). Since then, the annual number of reported rat bites has declined to fewer than 200, approximately 85 percent of which occur in congested areas where slum conditions and a high rate of infestation still exist. About 10 rat bites per 1 million persons are reported in large metropolitan areas, but there are probably several times that number of unreported bites (Clinton 1969).

In addition to their public health importance, rodents cause serious economic loss by eating and contaminating stored food and by damaging buildings through burrowing and gnawing.

Their gnawing of electrical and telephone wires can cause fires, power failure, or telephone interruptions. They also cause fires by taking into their nests such things as matches, lighted tobacco, and oily rags that ignite through spontaneous combustion. Solid waste is considered a major contributing factor in 30 to 50 percent of the fires in urban areas, and possibly 20 to 25 percent of the fires of undetermined origin are caused by rodents. When accumulations of solid waste (and consequently, rodents) have been reduced, structural fires have declined 50 percent (Walcott and Vincent 1975).

Although loss of foodstuffs and damage to structures by rodents are well known, accurate cost estimates are difficult to obtain (Jackson 1977). The figures cited in government reports are frequently derived from the unsubstantiated one rat/person ratio and a $10/rat/year "guestimate" of damage (Jackson 1977).

Lack of knowledge and of concern made the earliest attempts to control domestic rodents short-lived. The first organized efforts in the United States were made in the early part of this century, when the role of rodents in transmitting plague was recognized. During an outbreak of plague in San Francisco between 1904 and 1907, a comprehensive program aimed at eliminating food and harborage for rodents through public education, extermination, garbage and refuse management, and housing improvement was initiated. This program brought the epidemic under control in less than two years (Todd 1909).

Current control efforts in urban areas are based on an understanding of rodent population dynamics and limiting factors such as food and harborage. In urban areas garbage is the primary source of food for rodents, while vacant lots and deteriorated or abandoned buildings provide nesting sites. In a recent survey of 1,960 blocks in deteriorated neighborhoods in New York City, 11 to 35 percent of the premises examined were found to have improperly stored or exposed garbage. Unapproved refuse storage was found in 12 to 44 percent of the premises surveyed (New York City Department of Health 1977). Similar statistics can be cited for many other cities.

Poor sanitation and inefficient eradication measures also result in rodent infestations in various types of institutions and food establishments. A survey of 18,000 food establishments by the New York City Health Department showed that 87 percent had rodent or insect infestations, although 65 percent of them made use of the services of pest control operators (DuPree 1977).

The Federal Government has supported rat control programs in urban areas for a decade. Current federal assistance is justified for aesthetic and environmental reasons rather than for disease prevention. Annual appropriations for the program have exceeded $12 million, and 68 communities are currently involved (U.S. PHS 1979).

In the federally assisted program, inner-city blocks where rodent infestation is prevalent are designated by municipal governments as target areas for intensive rodenticide application, sanitation improvement, and health education. If rodents are reduced and sanitation is improved, the block is given maintenance status. If the improved status continues, federal assistance is withdrawn and the block becomes wholly a local responsibility again. About half of the 25,000 target-area blocks under the federal program are in maintenance status.

For the most part, the rodenticides used are anticoagulants, but several acutely toxic rodenticides may also be used to reduce the rodent population faster. Rodenticides often constitute only a part of the response to rodents, and the total amount of toxicant used by government agencies is difficult to determine.

The statistics in Table 3.22, compiled as part of the Committee's work, were taken from a survey of 28 urban areas receiving federal assistance. Project directors were asked to provide data on rodenticide use during the past three years (1976-1978), and data from the period of highest use were selected for the table.

Estimates of the cost of urban rodent control are elusive, but the annual cost of rodent control throughout the United States has been estimated at approximately $100 million (Brooks 1973).

There is no indication that human fatalities have occurred from the use of toxicants in federally assisted programs. Although some children have ingested some of the poisons, prompt medical attention has prevented adverse effects. Poisons applied by pest control operators and others have occasionally resulted in accidental ingestions and subsequent fatalities, however, and these incidents have often been highly publicized. Improper placement, lack of safety precautions, or incorrect use of rodenticides, coupled with lack of prompt medical attention, have been characteristic of such episodes. In addition, as noted earlier, some adults have used rodenticides in suicide attempts.

Fatalities of pets, especially dogs, have occurred with the use of some anticoagulants. Such incidents may occur from pet ingestion of the baits (especially paraffinized blocks), or from pet consumption of rodents that have eaten the poison. Accurate estimates of the number of pets killed are not available. Lack of restraints on pets, the large numbers of feral animals in cities, and lack of cooperation or attention by residents all contribute to this problem.

The genetic resistance of domestic rodents (both rats and mice) to the anticoagulant rodenticides was discovered in the United States early in the past decade (Jackson et al. 1971). Resistant populations of rats have been identified in 40 of the nearly 100 sites sampled (Jackson et al. 1975, Jackson and Ashton 1979). Where resistance has been encountered, the

TABLE 3.22 Summary of Annual Rodenticide Use in 28 Federally-assisted Urban Rat Control Programs

	Totals	Means[b]
No. of target area blocks	17,693	632
No. of reported rat bites in target area	121	4.3
Rodenticide Usage (pounds)[a]		
Fumigants (calcium cyanide)	204	34
Anticoagulants – grain bait	268,227	10,316
– paraffin blocks	33,600	1,867
– tracking powder	35	35
Acute rodenticides		
Vacor®	4,778	531
Zinc phosphide	14,328	1,592
Red squill	23,016	1,918
Norbromide	312	312
Antu	410	205

[a]Where concentrates were reported, the amounts were converted into finished baits for these tabulations. The data are from the period of highest use during 1976-1978.
[b]Calculations based only on projects using specified rodenticides.

SOURCE: Based on NRC Committee on Urban Pest Management Study of federally-assisted urban rat control programs.

older, acute rodenticides have been resorted to. Not all of the new "second-generation" anticoagulants, such as brodifacoum and bromadiolone, that are effective against resistant rodents and that are widely used elsewhere in the world are yet available in the United States (Jackson and Ashton 1979).

Education is an important aspect of rodent control in urban areas. The principles of rodent management and the fundamentals of good housekeeping and sanitation can be transmitted to urban residents through organized or informal educational programs. However, if the results of a survey conducted in Columbus, Ohio, are at all applicable generally, this transfer of information can be said to have already been accomplished (Sherer 1976). Most of the persons (86 to 99 percent) surveyed in Columbus knew the general conditions responsible for rodents; many (63 to 90 percent) could identify rodent signs; and a large number (70 to 75 percent) knew how to prevent rat infestation. What was lacking in Columbus was the involvement of individuals and neighborhood groups in organized programs directed toward initiating needed services and improving all aspects of the environment.

Urban rodent control programs should utilize a broad approach in which rodenticides supplement environmental improvement achieved through education and health and housing code enforcement.

Federally assisted urban rodent control programs focus on the rat, largely because of its size and psychological impact. The house mouse, however, also represents a significant but often unrecognized threat. Because of its ability to enter structures through small openings, minimal food requirements, limited movements, and secretive activity, the house mouse often lives in close proximity to people and their stored food, particularly in kitchens and warehouses. Direct and often insidious contamination of food supplies as well as transmission of disease pathogens and ectoparasites can result.

Statistics that separate infestations of rats from mice are scarce. FDA inspectors routinely record "rodent" droppings or hairs rather than attempting more precise identification. Many householders do not distinguish between young rats and mice and many believe that mice grow up to be rats! Since U.S. Public Health Service assistance to urban areas is for rats only, no data on mouse infestation has been developed by these urban programs.

Despite the lack of formal studies, pest control operators are well aware that mice are becoming increasingly difficult to control. This is related, in part, to the relatively higher dose of anticoagulant required to kill mice than rats. But more importantly, many PCO reports indicate that genetic resistance to anticoagulant rodenticides appears to be widespread. Laboratory determination of mice from all sectors of the country follow protocol established by the World Health Organization and a recent study in British Columbia (Cronin 1979). The phenomenon of pest resistance is so widespread in several European countries that conventional anticoagulants are no longer permitted in mouse control programs (Rennison 1971).

INTEGRATED MANAGEMENT OF PLANT PESTS

Although many attempts to manage plant pests are based almost exclusively on chemical pesticides, there also is evidence that nonchemical methods are being used alone or with chemicals to control plant pests (Frankie et al., in press, Olkowski et al. 1978a).

Street Trees in California

In 1970 a resident of Berkeley asked the city not to spray her tree with pesticide. Her request led to efforts by the University of California, Division of Biological Control to devise an alternative pest management strategy involving biological control of tree pests, primarily aphids. Ultimately, Berkeley became one of the first communities in California to participate in an urban IPM program for street trees. Street trees were chosen because of their visibility to the public, the public cost of maintaining them, and the accessibility of the public agencies charged with their care.

Basic research, applied research, and education are the three components of the program, which has been operating in Berkeley since 1970, San Jose since 1974, Palo Alto since 1975, and Modesto since 1976. In addition, the city of Davis and the Palo Alto School District participated in the program from 1976 to 1978. Approximately 447,000 trees are currently covered by the program, which originally focused on tree insect pests in the San Francisco Bay area. The program has now expanded, both in geographic range and in scope, to include tree diseases, indoor pests, and the pest problems involved in maintenance of state waterways.

The most important result of the program has been an average reduction in the number of insecticide treatments by 80 to 90 percent in the four cities. This has been accomplished because of successful monitoring programs; biological control of the linden, elm, and oak aphids; and education of the public and municipal employees. Pesticide treatments have largely been limited to either host-specific materials (such as *Bacillus thuringiensis* for lepidopterous pests) or to materials confined to a small area (such as Meta-Systox-R injections). Although no formal studies of changes in public opinion have been carried out, all of the cities have noted a reduction in the number of citizen requests for pesticide treatments, an outcome which suggests greater tolerance for pest problems. It thus appears that IPM can reduce public tree maintenance costs.

The program's major problem has been a shortage of funds for basic and applied research. Another problem has been the general unwillingness of maintenance personnel to adopt new pest management strategies out of fear of losing their jobs (Olkowski et al. 1978c).

Trees in Ohio

Research on the management of arthropod pests that attack woody ornamentals has been conducted in Ohio for several years. Some of this research has focused on integrated pest management, and the effort involving the bronze birch borer is reviewed below.

The bronze birch borer, *Agrilus anxius*, is a serious pest on birch trees from the East Coast to the Cascade Mountains (D.G. Nielson, Ohio Agricultural Research and Development Center, personal communication, 1979). Every year, numerous white-barked birches are killed by the boring activities of this beetle. Only senescent trees are infested in forest environments, but in urban areas birches of all ages are infested. The integrated approach to control of the borer has three major components:

1. Periodic watering and fertilization to maintain tree health and to preclude damage by aphids, leaf miners, and other pests in the spring, in recognition of the fact that vigorous trees are less frequently infested by the borer.

2. Pruning and destruction of dead branches, especially those that may have incipient borer infestations.

3. Selective use of lindane at specific times of the year to prevent recolonization of susceptible trees.

Although the program requires considerable effort, the effort is justified by the great value of these trees in urban environments. Furthermore, both arborists and homeowners can implement the program. Future research is expected to result in the identification of birch species that are unattractive to the borers or that are able to withstand borer infestation.

Herbivorous Pests Along California Freeways

The California Department of Transportation (CalTrans) engages in several programs to manage herbivorous arthropod pests along the state's freeways (Pinnock 1976). These programs have been largely inspired by the CalTrans "policy to eliminate the use of chemical pesticides, whenever practical to do so, that may be harmful to man and may also eliminate beneficial natural predators" (CalTrans 1975). Alternatives to pesticide use have been sought through cooperative urban pest management efforts with research personnel at the University of California, Berkeley.

Most of the programs involve the use of biological control agents, such as parasitic wasps, predators, and bacteria, against a variety of plant pests. Research has also been conducted on the use of a nonphytotoxic soft-soap solution against aphids. The soap, which is registered for use as a highway plant spray, is now widely used by CalTrans maintenance personnel. CalTrans has a policy of removing plants with signs of continued insect and/or disease problems.

Several aspects of the CalTrans programs are noteworthy. First, success of the programs has been largely due to the continued cooperation between CalTrans and the university and has provided incentive to become involved in other pest management schemes. Second, many of the programs are now conducted routinely by CalTrans personnel. Third, the programs also have negated the widespread use of pesticides.

Turfgrass Pests in Nebraska

The Committee on Urban Pest Management learned of only one IPM program on turfgrass, which began at the University of Nebraska early in 1979 (Gold 1978). The program is comprehensive in that it includes plant, arthropod, and plant disease components. In addition to the use of chemical and nonchemical methods, there will be considerable emphasis on the education of concerned parties.

Dutch Elm Disease

Although the Committee attempted to gather information on various IPM programs for dealing with Dutch elm disease, too little information was obtained to prepare an adequate review of this insect/fungus/elm system. It is known that Colorado, Minnesota, and New York are conducting research into IPM methods for dealing with the problem.

INTEGRATED MANAGEMENT OF COCKROACHES

In many areas of the United States, cockroaches are among the most prolific and disagreeable insects within and around urban structures (see Chapter 2). The pest control industry devotes much of its effort to treating cockroach infestations, and the National Pest Control Association estimates their economic importance as second only to termites (Cornwell 1976). Current insecticidal methods generally provide only short-term control within treated structures. Since several cockroach species regularly invade structures from outside sources, especially in the southern United States, reinfestation often occurs once the insecticide has dissipated.

Reliance on chemical methods of control has brought less than satisfactory results. The problems associated with using insecticides to control cockroaches include increased pest resistance, repellency, the costs of developing new and more powerful insecticides, the public health and ecological hazards of insecticide use, public awareness of the hazards, and government regulations. The severity of these problems indicates that alternative technologies and strategies are desperately needed. Furthermore, the realization that other methods may be more effective than insecticides in reducing cockroach populations to tolerable levels provides an additional reason for seeking viable alternatives.

Ecology and Behavior

Secluded places within urban dwellings provide dark and humid hiding and nesting sites that are similar to the microhabitats of tropical and subtropical regions where cockroaches have reached their greatest evolutionary development. Many studies have addressed the more obvious ecological factors in the survival and distribution of cockroaches (Cornwell 1968). Both the obvious and the more subtle ecological factors, as well as ways in which they can be used in a cockroach management program, are discussed by Ebeling (1980).

Very little information exists on the behavior and ecology of cockroaches in their nocturnal or diurnal environment. Recent technological advances in viewing devices have made it possible to observe cockroach activity at night, when domiciliary cockroaches are most active (Lewis 1979), but little use has been made of these devices, presumably because of their cost.

The effect of human habits on the behavior and ecology of the cockroach is frequently overlooked, while the influence that cockroaches (and other insects) have on human attitudes and behavior has only recently been explored (Frankie and Levenson 1978, Levenson and Frankie, in press). These studies indicate that people's attitudes and behavior toward cockroaches vary somewhat, depending on geographical location.

Cockroach Management in Texas

In 1972, Texas A&M University began work on an integrated pest management program to deal with cockroaches in middle-income residential areas of College Station and Houston. The goal of the program was to develop methods that professional pest control operators would be able to use without much difficulty (Piper and Frankie 1978, 1979).

The program began with intensive studies on the biology, behavior, and ecology of one of the state's most common pest species, *Periplaneta fuliginosa*. *Periplaneta americana* and *Blattella germanica* also were studied. These investigations formed the basis for the management scheme that was developed.

The management scheme consisted of up to five integrated tactics and tools and was tested on 11 residences. The first tactic was education of homeowners, who were given information on the biology, behavior, and ecology of the cockroaches in their homes. People were informed that their personal habits played an important role in the success or failure of the cockroach. The homeowners were then asked if they were willing to become involved with the management effort, since the second tactic, habitat management, greatly depended on them. Once they agreed to become involved, their homes were inspected for habitat modifications that would have to be made before other tactics could be employed. After the homeowners made these modifications, selected pesticides (e.g., boric acid), traps, or parasitic wasps (*Tetrastichus hagenowii*) were used (Piper and Frankie 1978, 1979).

Initial meetings with homeowners focused on determining the number of cockroaches they could tolerate in the house and led to the development of an aesthetic injury level (AIL) for each residence. The tolerable level was used, in part, as a basis for evaluating the efficacy of the program.

In general, the greatest control was observed when all of the tactics were employed. In some cases, however, not all tactics were needed. Judgment of the program's effectiveness was based on how often the AIL was surpassed, as well as on homeowner satisfaction. In residences in which most of the tactics were used in combination, the AIL was rarely exceeded during the study period. When the level was exceeded, a reexamination of the premises usually revealed that additional habitat modifications were required. Homeowner satisfaction was high in almost all cases, even when the AIL was surpassed.

Although the program showed promise as an effective way of dealing with cockroaches, several aspects require further study. First, no assessment was made of the program's efficacy in reducing cockroach populations outdoors. This is important because outdoor populations generally provide the reservoir for indoor populations. Second, no effort was made to assess the impact of the program on human health or on other nontarget organisms. Third, implementation of the program by professional pest control operators (PCO) was not attempted. Fourth, the number of homeowners who continued the program on their own or sought the services of PCOs after the study period is not known.

Cockroach Management in New Jersey

A program for the management of German cockroaches in low-income private housing and inner-city restaurants and institutions in Trenton was initiated in 1973 (A.P. Gupta, Rutgers University, personal communication, 1979). The overall objectives of the program were (1) to develop safer methods of chemical control of the German cockroach in homes, restaurants, hospitals, and nursing homes over the longest possible period of time through a single application of the least toxic chemical; (2) to demonstrate the importance of sanitation and education in effective control; (3) to develop a suitable mechanical trap to be used with food or juvenile-hormone-treated baits, natural or synthetic pheromones, or related compounds in conjunction with chemical control methods; (4) to study the effects of juvenile hormone analogs (JHAs) on the reproductive physiology and population dynamics of the German cockroach; and (5) to monitor cholinesterase depression levels in the residents of treated homes, hospitals, and nursing homes and to monitor insecticide residue levels in the homes themselves.

The program was developed as a university research project and required the interdisciplinary cooperation of a community organization and two local public health agencies (Gupta et al. 1975).

Initially, 56 homes were surveyed to determine existing infestation levels, sanitary conditions, and the willingness of residents to cooperate (Gupta et al. 1973). Of the homes surveyed, 24 were selected for the study. Treatment consisted of a single application of moderately toxic (Dursban®, diazinon) or low-toxic (resmethrin) insecticides in combination with boric acid. Visual counts of cockroaches using a flushing agent (synthetic pyrethroid) were made at 2-week intervals for 9 months to determine the efficacy of the treatment (Gupta et al. 1973).

In general, poorer control was found in homes or restaurants with inadequate sanitation. In homes with good sanitation, a 98 percent level of control was achieved over the 9-month period. Resmethrin-boric acid treatments proved most desirable for use in homes, restaurants, hospitals, and nursing homes. In three restaurants, the combination achieved a 77 to 98 percent level of control within 6 months and an 81 percent level of control after 9 months. In the housing units surveyed, overall insecticide use was reduced 15-fold compared with other commercial operations in the area (A.P. Gupta, Rutgers University, personal communication, 1979).

JHAs showed promise as an integrated tool for controlling cockroaches. They inhibited mating in treated females, produced malformed internal reproductive organs, and rendered females unable to produce oothecae. The ultimate goal of this aspect of the program is to formulate (a) a male pheromone bait to lure virgin females and (b) a JHA that will inhibit mating and cause morphogenetic abnormalities (A.P. Gupta, Rutgers University, personal communication, 1979).

Only the first two objectives of the program were realized before it was unexpectedly discontinued in 1975. It is significant, however, that a high degree of control was achieved for 9 months from a single application of insecticide when used in conjunction with accepted sanitary practices. The program is to be renewed in 1980 and will cover a longer period of time and focus on additional pests.

Cockroach Management in Maryland

A program for the management of German cockroaches in 500 low-income public housing units in Baltimore was initiated in 1976 (F.E. Wood, University of Maryland, personal communication, 1979). The program stresses education of tenants, accurate record keeping, and selective use of insecticides on a focus-treatment basis after initial treatment. In addition to tenants, other participants include commercial pest control operators (PCOs), social workers, health department inspectors, family counseling groups, extension service food and nutrition aides, school health aides, and other community workers.

At the outset, there is a group meeting for tenants, management, PCOs, and others to discuss problems and solutions. Initially, the PCO surveys the apartments to evaluate existing cockroach populations and household sanitation. A special effort is made to relay the findings to the tenants in a friendly and informative manner, and advice on house-cleaning, sanitation practices, and pest harborage sites is provided by social workers. Depending on the initial level of infestation, the PCO treats each unit with chlorpyrifos and silica gel with pyrethrin. No fog or spray formulations are used. Units with very high pest populations ("focus units") are surveyed four weeks after the initial treatment, and treatment is repeated. On the third and subsequent visits, the procedures include survey, evaluation, and dusting in strategic places. Where high populations of cockroaches persist, an analysis of the situation is made and remedial sanitation measures are enforced. The same PCO conducts all survey and treatment activities and advises tenants on sanitation practices to reduce the need for insecticides.

On the basis of tenant purchases of over-the-counter insecticides, insecticide use decreased approximately sevenfold within six months. Insecticide expenditures also dropped from about \$25 to \$3.50 per tenant. Surveys indicated that 91 percent of the tenants felt that cockroach control had improved and that the pest management scheme was acceptable.

On the basis of these tests, the Baltimore Housing and Community Development Agency is planning to adopt this program for 38 housing projects beginning in February 1980 (F.E. Wood, University of Maryland, personal communication, 1979).

Cockroach Management in California

In the early 1970s a program for resolving cockroach problems at the University of California in Berkeley was developed through cooperative efforts of the university's academic and administrative staff (G.W. Frankie, University of California, personal communication, 1979). The Berkeley campus has more than 400 buildings that contain 4,200 living units (apartments and dormitory rooms), 35 libraries, 14 food service areas (commissary and cafeterias), several hundred animal rooms, and 3,000 laboratories.

A variety of control techniques, including insecticide spraying, parasites, trapping, habitat modifications, and education, were used in efforts directed at the two species of cockroaches, German and brown-banded, that are the major domicilliary pests on campus. The techniques were selectively applied to meet requirements for effectiveness and efficiency, to meet regulatory requirements, and to meet funding and staffing limitations.

The program reduced the amount of liquid insecticides used in food service areas by 90 percent, in student housing by 94 percent, and in research facilities by 99 percent. The program has achieved the following results:

1. Harmful effects on nontarget organisms, research animals, and human beings are avoided.
2. Academic personnel have expressed strong support for the techniques as being suitable to their needs.
3. Five other institutions have adopted parts of the program.

General Discussion of Cockroach Management Programs

All of the IPM programs discussed above have a number of features in common. Each program has an ecological focus that is essential for an integrated effort. In each program (1) pest situations were treated individually; (2) account was taken of the role that personal habits play in determining the environment of cockroach populations, and efforts were made to modify habitats; (3) only insecticides of low to moderate toxicity were used; (4) overall use of insecticides was lower than in insecticide-based programs; (5) pest populations were directly monitored through inspection of premises and indirectly by educating residents to pay closer attention to the degree of infestation; and (6) the area of infestation was considered, not just a single premise.

Education is one of the most important aspects of IPM, but it has been neglected in past efforts to control cockroaches. Educating the public about cockroaches deserves considerably more attention and emphasis.

Although scientists and other academic personnel developed these programs, there is no reason to believe that PCOs and paraprofessionals with some understanding of ecology could not develop similar ones.

INTEGRATED MANAGEMENT OF FLIES ON MACKINAC ISLAND

Mackinac Island is one of the most attractive tourist sites in Michigan. Located five miles from the mainland in the Strait of Mackinac, the island encompasses 2,465 acres and is visited each summer by more than 750,000 people (E. Peterson, Mackinac Island State Park, personal communication, 1978).

Perhaps the best-known feature of this popular tourist attraction is the use of horse-drawn carriages. The presence of a large number of horses, however, results in large accumulations of dung which provide breeding sites for pestiferous flies. Despite efforts to dispose of the dung, it accumulates quickly enough to allow propagation of large numbers of the housefly (*Musca domestica*) and the biting stable fly (*Stomxys calcitrans*). These flies typically breed in disturbed manure mixed with straw and urine in and around stables and barns.

A program to control adult flies has been conducted on the island for many years (Hoopingarner et al. 1966). The chlorinated hydrocarbon DDT was first used in 1945 and gave excellent control. When resistance to this compound and other chlorinated hydrocarbons became evident, malathion sprays were substituted until 1964, when resistance again became a problem (Hoopingarner and Krause 1968). Dimethoate was then used, and a few years ago the spray schedule was intensified so that all infested areas were treated at least once every 7 to 10 days. As of 1978, however, the fly population appeared to be building up resistance to dimethoate (R. McCreedy, Mackinac Island State Park, personal communication, 1978).

In addition, a second pest problem has developed in recent years on the island. An outbreak of the European fruit lecanium scale complex (*Lecanium corni*) has seriously affected many of the shade and fruit trees located in or near the park (M.K. Kennedy, Michigan State University, personal communication, 1977). Many branches have died, and there has been a general decline in the vigor of infested trees. Under such stressful conditions, trees are also more vulnerable to attack by other insects (especially wood-boring beetles) and disease. Preliminary observations indicate that the dramatic increase in the scale is associated with weekly applications of dimethoate for control of flies. Dimethoate appears to have eliminated the scale's natural enemies (parasites and predators), allowing it to increase to damaging levels.

The fly control program has thus created three serious problems: (1) increasing insecticide resistance in a pest population; (2) a secondary pest outbreak; and (3) increased application of a toxic chemical with potential development of human exposure.

In 1978 an integrated pest management program was initiated that included physical/chemical methods (poison fly traps) and sanitary measures (curtailment of breeding sites and dung composting). Insecticide use, meanwhile, was severely curtailed. This program resulted in a 40 to 50 percent reduction in the fly populations during 1978 and 1979 (Kennedy and Merritt 1980). Furthermore, there was a dramatic decrease in *Lecanium*, at least in the downtown area. The program also eliminated human exposure to insecticides and reduced contamination of the environment.

The program will be continued in 1980, at which time university researchers will train state park and city workers in control techniques, especially sanitation methods. It is expected that eventually the program will be carried on by nonuniversity personnel.

The IPM program, however, has also resulted in new problems. There have been some increases in the number of yellowjackets, mosquitoes, tabanids, and lilac leaf miners. Although the program has not directly caused these increases, it has apparently weakened natural controls on the other pests. This has led to a variety of social and political problems that university research personnel have had to address through educational efforts (R.W. Merritt, Michigan State Univeristy, personal communication, 1979).

MOSQUITO CONTROL

Approaches to control of mosquitoes and of the human diseases transmitted by mosquitoes have changed substantially during the past century. The success of Gorgas in controlling yellow fever and malaria during the building of the Panama Canal (1904 to 1907), and the subsequent eradication of *Aedes aegypti* from Brazil in the 1930s focused control efforts on source reduction, that is, elimination of breeding sites. This approach culminated in the monumental efforts of the Tennessee Valley Authority (1935 to 1945) to control mosquitoes in the area by managing water levels in reservoirs, encouraging installation of household screening, and selective larviciding—in short, by conducting an effective IPM program.

The advent of DDT in 1939 provided an alternative to source reduction. The new chemical was cheap, was highly effective against both larvae and adults, persisted for 6 months to a year on the walls and ceilings of all sorts of dwellings, and seemed to be harmless to human beings. To many, the opportunity proved irresistible. In the Latina Province of Italy, for example, the town of Missiroli organized a residual house-spraying program that eliminated malaria between 1944 and 1947. Logan of the Rockefeller Foundation eradicated malaria in Sardinia between 1945 and 1951 by using DDT to control mosquitoes. The World Health Assembly endorsed malaria eradication in 1955, and by 1960 residual spraying of dwellings with DDT—and to a lesser extent with BHC and dieldrin—had reached global proportions.

This simple strategy of insect eradication was doomed to failure, however, for it resulted in intensive selection of insecticide-resistant races of *Anopheles*. *A. sacharovi* in Greece was the first to develop widespread resistance to DDT. Dieldrin, chlordane, and BHC were substituted, but strong resistance to dieldrin appeared in 1952 and malaria, which had almost completely ceased to afflict the Greeks, reappeared in epidemic proportions in 1956. This chain of events has recurred many times throughout the world as major malaria vectors have developed virtual immunity to insecticides: *A. sundaicus* in Indonesia to DDT in 1954, *A. gambiae* in West Africa to BHC and dieldrin in 1955, *A. stephensi* in the Persian Gulf to DDT in 1955, *A. albimanus* in Central America to DDT in 1958, *A. pharoensis* in Egypt to DDT and dieldrin in 1959, and *A. culicifacies* in India to DDT in 1960 (Brown and Pal 1971).

By 1976 there were 42 species of *Anopheles* resistant to chemical insecticides, including 41 resistant to dieldrin and 24 resistant to DDT (WHO 1976). Substitution of malathion and fenitrothion resulted in the development of resistance in *A. messae* in Romania, *A. sacharovi* and *A. hyrcanus* in Turkey, *A. sinensis* in Korea and Japan, and *A. culicifacies* in India. *A. albimanus* is now resistant to DDT, dieldrin, BHC, organophosphate insecticides, and to the carbamate propoxur in El Salvador, Guatemala, Honduras, and Nicaragua. Major epidemics of malaria have recurred in Sri Lanka, India, and Pakistan. Resistance of anopheline vectors to insecticides had become a major problem in areas with more than 256 million inhabitants. As a result, the World Health Organization since 1976 has tended to speak of malaria "control" instead of malaria "eradication" (WHO 1976).

The history of other mosquito vectors is much the same. *Culex pipiens fatigans* had become markedly resistant to DDT in Brazil by 1952 and also became resistant elsewhere in South America to DDT, BHC, and dieldrin. Failure to control *Culex pipiens* with BHC was experienced in California in 1951. Subsequently, *C. pipiens* achieved resistance in India to DDT in 1952 and to BHC and dieldrin in 1954. Both DDT and chlordane failed to control *C. pipiens* in the Ryukyu Islands in 1954, and there was cross-resistance to BHC and dieldrin. Malathion was substituted in 1959, and by 1967 there was pronounced resistance to malathion with cross-resistance to fenthion. In Africa, dieldrin and BHC resistance in *Culex pipiens* was detected in 1958. Resistance to malathion and diazinon became pronounced in 1960.

Culex tarsalis, the vector of western equine encephalitis, became resistant to DDT in 1947, to BHC in 1949, and by 1951 was also resistant to aldrin, heptachlor, and toxaphene. When the organochlorines were replaced by the organophosphate insecticides parathion, EPN, and malathion, resistance to malathion appeared after 2 years. Within a decade, generalized organophosphate resistance was present in this mosquito almost everywhere in the Central Valley of California. *Culex tarsalis* and *Aedes nigromaculis* are now resistant to almost every available insecticide.

Aedes aegypti, the mosquito vector of yellow fever and dengue, was initially highly susceptible to DDT and became the target of widespread insecticide programs. An *A. aegypti* eradication campaign in the Americas was initially highly effective in most of South and Central America, but by 1950, DDT- and dieldrin-resistant species had begun to appear in a number of countries. A highly touted campaign to eradicate *Aedes aegypti* in the United States was quietly abandoned in 1968, after the expenditure of $80 million, when it was determined that the mosquito's resistance to DDT was widespread (WHO 1976). At present, *A. aegypti* is resistant to DDT, BHC, and dieldrin, and to the organophosphate insecticides malathion, fenthion, fenitrothion, and diazinon.

By 1976, 83 resistant species of mosquitoes had been noted (Georghiou and Taylor 1978), most of them showing multiple resistance to a variety of insecticides. As a result, spreading kerosene or diesel oil on waters where mosquito larvae flourish has once again become the favorite method of control in many areas. Exclusively insecticidal control of mosquitoes now has a dim future, and new programs must employ IPM so that suppression by a variety of means may serve to reduce the development of insecticide resistance.

Mosquito control programs have returned to source reduction as the basic technique around which other components of IPM are arrayed. Source reduction now involves deliberate modification of the aquatic environment to render it unsuitable for mosquito production. Source reduction efforts include total removal of standing water by draining, ditching, diking, and filling; manipulation of water levels; and alterations of the aquatic habitat by mowing, clearing, or otherwise manipulating plant species (Metcalf 1975b). The intricacies of integrated pest management of mosquitoes are too complex to detail here, particularly since they must be tailored to fit each locality and vector system, but indispensable components include natural control by predatory fish such as *Gambusia affinis,* by invertebrate predators such as the larvae of *Toxorhynchites,* and by such diseases as the parasitic nematode *Reesimermis nielseni*, the microsporidam *Nosema* spp., and *Bacilius sphaericus*. Other important components include mosquito-proofing of human dwellings (especially by proper screening) and construction of adequate sewage disposal facilities. Selective use of insecticides—oils and granular applications for larvae, pyrethroid sprays for adults—also have a place in such programs (Metcalf 1975b).

MANAGEMENT OF PESTS IN FOOD-HANDLING ESTABLISHMENTS

Food passes through many channels and processes from production to consumption. Throughout this system, food is subject to attack by more than 200 rodents, birds, insects, and mites (Gorham 1975). Numerous state and federal laws have been passed to protect the integrity of food, most notably the Federal Food, Drug, and Cosmetic Act (FFDCA) (21 U.S.C. 301 et seq.), the Federal Meat Inspection Act (19 U.S.C. 1306; 21 U.S.C. 601-623, 641-645, 661, 671-680, 691), and the Poultry Products Inspection Act (21 U.S.C. 451-470).

Gorham (1974) has reviewed in detail the implications for human health of filth in food. "Filth" includes contamination by animals—rodents, insects, or birds—as well as objectionable matter resulting from unsanitary conditions. Food pests may adversely affect human beings in numerous ways, but under the Federal Food, Drug, and Cosmetic Act it is not necessary to demonstrate a cause-and-effect relationship to establish a basis for enforcement action. In fact, enforcement actions under the FFDCA are commonly based solely upon evidence of extraneous materials in food products or on evidence that storage or processing was done under unsanitary conditions. Compliance with the FFDCA requires, among other things, that food must not be prepared, packed, or held under unsanitary conditions and that food manufacturers must use manufacturing processes currently deemed to be satisfactory.

Food products can be contaminated by pesticides as well as by pests. The U.S. Food and Drug Administration (FDA) has statutory responsibilities for control of pesticide and other chemical residues in food that include:

1. Monitoring food for chemical residues and, when illegal residues are found, initiating regulatory action;
2. Gathering information on incidence and amounts of chemical residues in food to permit evaluation of the effectiveness of federal regulations, identification of emerging problems, establishment of allow-

able tolerances for chemicals in food, and provision of information to EPA for that agency's decisions on pesticide registrations;

3. Informing the public, Congress, industry, and other concerned groups about chemical residues in the diet.

Despite high levels of sanitation and product care in the food industry, pest management with or without chemicals cannot be expected to produce food and feeds that are free of all foreign matter. Educating the public about this reality is an important element of any pest management program, since such problems are likely to increase because of integrated pest management programs. While pest infestations may not reach levels that are economically important to the grower, they may alarm consumers or retailers. Some reasonable understanding is necessary in order to avoid wasting food or using pesticides when only trimming and washing may be needed.

Pests of Stored Products

Many methods are employed in the management of pests that infest stored products. They range from one-time fumigation to highly sophisticated integrated programs in which several methods complement one another. Pederson and Mills (1978) observed that in no type of pest control has a greater variety of techniques been used than in control of stored-product pests. The methods include sanitation, temperature variation, moisture control, fumigants, aeration, protectants, insecticides, gases, radiation, pheromones, insect growth regulators, insect pathogens, predators, insect-resistant packaging, and resistant food varieties. The authors also note harvesting, storage, and management techniques that affect the degree of pest damage. Development and exploitation of these methods are being carried out in U.S. Department of Agriculture laboratories and by larger firms in the food industry.

Pests in Transportation Vehicles

Food in rail cars and trucks may be attacked by pests that breed in wastes from previous cargo. Cogburn (1973) found 29 species of stored-product pests associated with about 60 percent of the boxcars at Gulf Coast ports; during summer months, food in up to 90 percent of the boxcars may be infested. This problem was recently addressed by a task force of specialists from the food industry, the government, and the railroads (Henderson and Meister 1977). Their principal recommendation was that infested rail cars should be fumigated by trained and experienced personnel. Details are provided on how to conduct inspections and which factors should be considered in deciding if fumigation is needed. The task force also mentioned the use of appropriate residual sprays to create a barrier between the commodity and the pests in the fabric of the vehicle.

Pests in Food Warehouses

Inspections at food warehouses are authorized by the FDA and local authorities. Numerous pests and their filth in food warehouses may also be transported to retail food stores. Inadequate design of warehouses, loading docks, and staircases; improper reception, storage, and rotation of stock; and inadequate control strategies all hamper pest management in food warehouses.

Although an integrated pest management program for food distribution warehouses has been developed by the pest control industry (NPCA 1979a), additional educational input on the basics of warehouse pest management is needed. Results obtained by Cogburn (1973) in 4 Gulf Coast warehouses indicate that although pest management strategies reduce postharvest pests in warehouses, the pests are not completely eliminated. Warehouse owners, managers, and employees all need training in the diverse components of and necessity for pest management in food warehouses.

Pests in Retail Food Stores

Retail food stores in the inner city present extremely difficult pest management problems. Many such businesses operate in cramped quarters where sanitation is difficult to achieve. Merchandise is delivered by a variety of suppliers, so that rejection of infested commodities is rarely practical. Customers returning empty containers or unsatisfactory merchandise may introduce pests. Economy and convenience to customers may be viewed as more important than the practice of integrated pest management (NPCA 1979b).

In suburban and less-congested city areas, pest problems are less severe. Investment is greater, and so are the incentives to avoid infestation. In the Washington, D.C. area, for example, a leading supermarket firm that recently opened a new retail outlet of 40,000 ft^2 spent $1 million for construction, land, and fees; $1.2 million for equipment and fixtures; and $360,000 for opening day merchandise (see Washington Post, November 23, 1977). Such stores sell up to $12 million worth of merchandise each year and employ about 100 persons. All of this investment can be endangered by the presence of pests.

"Integrated Pest Management in Retail Food Stores" is the subject of a recent report of the Food Protection and Sanitation Committee of the NPCA (1979b). The document reviews pest management practices of the commercial pest control industry and notes that well-trained and experienced personnel are required for IPM programs. These programs are increasingly being used by the larger grocery firms, but education, motivation, and improved incentives are required to make the long-term benefits of IPM attractive to a majority of those in the retail food trade.

Pests in the Food Service Industry

Pest management is a moral, legal, and economic necessity in restaurants, cafeterias, and other places where meals, snacks, and drinks are served. Restaurant industry officials report that 145 million customers are served daily in 541,000 restaurants, cafeterias, drive-ins, snack stands, department stores, drugstores, coffee shops, employee cafeterias, schools, hospitals, hotels, and other establishments (Bower and Davis 1976). The building in which the business is housed, the furnishings, and the equipment for food preparation and service all provide harborage for a variety of pests attracted by the odors released during food storage, preparation, service, and disposal. Pests may also get into the establishment in produce or dry commodities, in packaging, or in nonfood items such as linens or paper goods. Pests may also enter in the clothing or personal effects of workers or customers.

Pests in food establishments present potentially serious health hazards through contamination of food. They also add to the costs of owners, since foods that have been adulterated or fouled by pests must be disposed of.

The presence of pests also has serious effects on customers. A recent national survey conducted for the National Restaurant Association (1977) asked people to identify the characteristics they considered important in choosing or returning to a particular restaurant. Cleanliness ranked first in selecting a quick-service or moderate-service restaurant; in full-service restaurants cleanliness ranked second only to food quality and preparation.

Recommendations to the food service industry emphasize that it is necessary to prevent pest contamination without introducing contamination by pesticides (National Restaurant Association 1979). Pest prevention can be accomplished by effective management of a food establishment's physical aspects, as follows:

1. Location and exterior surroundings. Location and exterior surroundings can encourage the proliferation of pests. To the extent that the

establishment is situated in a clean and pest-free environment, the problem is minimized.

2. Structural integrity. The maintenance of structural integrity, including the use of screens and the elimination of holes and other openings, can help to prevent some infestation.

3. Equipment. Pest problems can be reduced if equipment is designed for easy cleaning and with no room for pests.

Certain practices, if carried out routinely, can greatly reduce or eliminate the likelihood of pest infestation. These include careful inspection of incoming supplies, clean and dry storage of perishable foodstuffs, frequent cleaning of the premises, and careful disposal of garbage and waste water.

Hasty and uninformed pest control programs in food-handling establishments can lead to contamination of food with pesticides. Since pests are more visible than pesticide residues, careless or unnecessary pesticide treatments may occur.

In order to meet public expectations and legal requirements to serve food that is essentially free of contamination by pests or pesticides, the food service industry has developed a number of successful pest management techniques, many of which are based on IPM principles. Although the importance of owner pride, product stewardship, and public responsibility as incentives for successful pest management programs cannot be ignored, it is sometimes necessary to resort to more forceful measures, such as threats of fines or imprisonment of corporate officials.

REFERENCES

Ashton, A.D. and W.B. Jackson (1979) Field testing of rodenticides in resistant-rat area of Chicago. Pest Control 47(8):14-16.

Aswood-Smith, M.J., J. Trevino, and R. Ring (1972) Mutagenicity of dichlorvos. Nature 240:418-420.

Barker, R.L. (1958) Notes on some ecological effects of DDT sprayed on elms. Journal of Wildlife Management 22:269-274.

Bower, W.F. and A.S. Davis (1976) The federal food service program. Journal of Milk and Food Technology 39:128-131.

Brooks, J.E. (1973) A review of commensal rodents and their control. Critical Reviews in Environmental Control 3(4):405-453.

Brown, A.W.A and R. Pal (1971) Insecticide Resistance in Arthropods. WHO Monograph Series 38. Geneva: World Health Organization.

Bush, G.L., R.W. Neck, and G.B. Kitto (1976) Screwworm eradication: Inadvertent selection for noncompetitive ecotypes during mass rearing. Science 193:491-493.

California Department of Food and Agriculture (1979a) 1978 Pesticide Use Report, Annual. Sacramento: California Department of Food and Agriculture.

California Department of Food and Agriculture (1979b) Pesticide Use Report by Commodities, 1978. Sacramento: California Department of Food and Agriculture.

California Department of Transportation (1975) Manual of Roadside Maintenance. Sacramento: California Department of Transportation.

Chamberlain, W.F. (1979) Methoprene and the flea. Pest Control 47(6):22-26.

Chappelka, R. (1979) Two-year Summary Report of Possible Vacor Incidents, April 5, 1977 to April 4, 1979. Philadelphia: Rohm and Haas Company.

Clausen, C.P., ed. (1978) Introduced Parasites and Predators of Arthropod Pests and Weeds: A World Review. U.S. Department of Agriculture Handbook 480. Washington, D.C.: U.S. Government Printing Office.

Clinton, J.M. (1969) Rats in urban America. Public Health Reports 84(1):1-7.

Cochran, D.G. (1976) Disjunction types and their frequencies in two heterozygous reciprocal translocations of *Blattella germanica* (L.). Chromosoma (Berlin) 59:129-135.

Cochran, D.G. (1977) Patterns of disjunction frequencies in heterozygous reciprocal translocations from the German cockroach. Chromosoma (Berlin) 62:191-198.

Cochran, D.G. (1979) A genetic determination of insemination frequency and sperm precedence in the German cockroach. Entomology Experiments and Applications 26(3):259-266.

Cochran, D.G. and M.H. Ross (1969) Chromosome identification in the German cockroach. Journal of Heredity 60:87-92.

Cochran, D.G. and M.H. Ross (1974) Cytology and genetics of T(9;11) in the German cockroach, and its relationship to other chromosome 9 traits. Canadian Journal of Genetics and Cytology 16:639-649.

Cochran, D.G. and M.H. Ross (1977a) Cytology and genetics of a stable ring-of-six translocation in the German cockroach. Journal of Heredity 68:172-178.

Cochran, D.G. and M.H. Ross (1977b) Cytogenetics of T(11;12) a reciprocal translocation in the German cockroach. Journal of Heredity 68:379-382.

Cogburn, R.R. (1973) Stored-product insect populations in port warehouses of the Gulf Coast. Environmental Entomology 2(3):401-407.

Colorado Department of Agriculture (1979) Pesticide Use Data 1977-78. (Submitted by R.B. Turner, Chief, Pesticide Section, to L.C. Wallace, Staff Officer, NRC Committee on Urban Pest Management, August 25, 1979.) (Unpublished)

Cornwell, P.B. (1968) The Cockroach. Vol. I. London: Hutchinson.

Cornwell, P.B. (1976) The Cockroach. Volume II. New York: St. Martin's Press.

Cronin, D.E. (1979) Warfarin Resistance in House Mice (*Mus musculus* L.). M.P.M. Professional Paper. Vancouver, British Columbia: Simon Fraser University.

Davies, J.E. (1973) Pesticide residues in man. Pages 313-333, Environmental Pollution by Pesticides, edited by C.A. Edwards. London: Plenum Publishing Company.

Davies, J.E. and W.E. Edmundson (1972) Epidemiology of DDT. New York: Futura Publications.

Davies, J.E., W.E. Edmundson, A. Raffonelli, and C. Morgade (1972) The role of social class in human pesticide pollution. American Journal of Epidemiology 96(5):334-341.

Davies, J.E., W.E. Edmundson, and A. Raffonelli (1975) The role of house dust in human DDT pollution. American Journal of Public Health 65(1):53-57.

Davis, D.E. (1950) The rat population in New York, 1949. American Journal of Hygiene 52(2):147-152.

Davis, B.N.K. (1978) Urbanization and the diversity of insects. Pages 126-138, Diversity of Insect Faunas, edited by L.A. Mound and N. Waloff. Royal Entomological Society Symposium 9. London: Blackwell.

Davis, D.E. and W.T. Fales (1949) The distribution of rats in Baltimore, Maryland. American Journal of Hygiene 49(3):247-254.

Davis, D.E. and W.T. Fales (1950) The rat population of Baltimore, 1949. American Journal of Hygiene 52:143-146.

DeBach, P. and K.S. Hagen (1964) Manipulation of entomophagous species. Pages 429-458, Biological Control of Insect Pests and Weeds, edited by P. DeBach. London: Chapman and Hall.

Doutt, R.L. (1961) The dimensions of endemism. Annals of the Entomological Society of America 54:46-53.

DuPree, R.A. (1977) New York problems and one solution. Pages 4-6, Proceedings of the 1977 Seminar on Cockroach Control. Albany, New York: New York State Department of Health.

Ebeling, W. (1980) Ecological Aspects of Cockroach Management. U.S. Food and Drug Administration Technical Bulletin, edited by A.R. Gorham. Washington, D.C.: U.S. Public Health Service.

Ehler, L.E. and G.W. Frankie (1979a) Arthropod fauna of live oak in urban and natural stands in Texas. II. Characteristics of the mite fauna (Acari). Journal of the Kansas Entomological Society 52:86-92.

Ehler, L.E. and G.W. Frankie (1979b) Arthropod fauna of live oak in urban and natural stands in Texas. III. Oribatid mite fauna (Acari). Journal of the Kansas Entomological Society 52:344-348.

Felt, E.P. (1905) Insects affecting park and woodland trees. New York State Museum Memoir 8:46-48.

Force, D.C. (1974) Ecology of insect host-parasitoid communities. Science 184:624-632.

Frankie, G.W., D.L. Morgan, M.J. Gaylor, J.G. Benskin, W.E. Clark, H.C. Reed, and P.J. Hamman (1977) The mealy oak gall on ornamental live oak in Texas. Texas Agricultural Experiment Station, Miscellaneous Publication 1315.

Frankie, G.W. and L.E. Ehler (1978) Ecology of insects in urban environments. Annual Review of Entomology 23:367-387.

Frankie, G.W. and C.S. Koehler, eds. (1978) Perspectives in Urban Entomology. New York: Academic Press, Inc.

Frankie, G.W. and H. Levenson (1978) Insect problems and insecticide use: Public opinion, information and behavior. Pages 359-399, Perspectives in Urban Entomology, edited by G.W. Frankie and C.S. Koehler, New York: Academic Press, Inc.

Frankie, G.W., H. Levenson, and S. Mandel (In Press) Homeowner pesticide useage in three U.S. metropolitan areas (tentative title). Hilgardia.

Frankie, G.W. and C. Magowan (In Press) A preliminary study of attitudes and practices of pest control operators toward pests and pesticides in selected urban areas of California and New Jersey. Hilgardia.

Gates, R.L. (1957) Diphacin, new anticoagulant rodenticide. Pest Control 25(8):14 and 16.

Georghiou, G.P. and C.E. Taylor (1978) Pesticide resistance as an evolutionary phenomenon. Pages 759-785, Proceedings of XV International Conference of Entomology, Washington, D.C., April 19-27, 1976.

Gerhold, H.D., E.J. Schreiner, R.E. McDermott, and J.A. Winieski (1966) Breeding Pest Resistant Trees. New York: Pergamon Press.

Glass, E.H. (1975) Integrated Pest Management: Rationale, Potential, Needs and Implementation. Entomological Society of America Publication No. 75-2. College Park, Md.: Entomological Society of America.

Gold, R.E. (1978) Comprehensive Nebraska Proposal for Integrated Turfgrass Pest Management. Lincoln, Nebraska: University of Nebraska.

Gorham, J.R. (1974) Filth in foods: Implications for health. Journal of Milk and Food Technology 38:409-418.

Gorham, J.R. (1975) Arthropod Pests of the Food Industry: A List and Bibliography of Identification Aids (Revised). Cooperative Economic Insect Report. Washington, D.C.: U.S. Department of Agriculture.

Granovsky, T.A. and G.W. Frankie (In Press) Texas pest control operators' attitudes and reactions toward pests and pesticides in Dallas, Texas, summer 1978. Hilgardia.

Gupta, A.P., Y.T. Das, J.R. Trout, W.R. Gusciora, D.S. Adam, and G.J. Bordash (1973) Effectiveness of spray-dust-bait combinations and the importance of sanitation in the control of German cockroaches in an inner-city area. Pest Control 41:20-26, 58-62.

Gupta, A.P., Y.T. Das, W.R. Gusciora, D.S. Adam, and L. Jangowsky (1975) Effectiveness of 3 spray-dust combinations and the significance of "correction treatment" and community education in the control of German cockroaches in an inner-city area. Pest Control 43(7):28, 30-33.

Hagen, K.S., E.F. Sawall, Jr., and R.L. Tassan (1971) The use of food sprays to increase effectiveness of entomophagous insects. Pages 59-81, Proceedings of the Tall Timbers Conference on Ecology, Animal Control, and Habitat Management, Volume 2.

Hayes, W.J., Jr. and W.K. Vaughn (1977) Mortality from pesticides in the United States in 1973 and 1974. Toxicology and Applied Pharmacology 42:235-252.

Heath, D.F. (1961) Organophosphorus Poisons. New York: Pergamon Press.

Henderson, L.E. and H.E. Meister, Jr. (1977) Guidelines for Pest Control in Railcars for Food Transportation. USDA Program Aid 1178. Washington, D.C.: U.S. Department of Agriculture.

Hoopingarner, R.A., G.E. Guyer, and D.H. Krause (1966) The Mackinac Island fly problem. I. History of insecticide use and characterization of resistance. Michigan Agricultural Experiment Station Quarterly Bulletin 48:559-564.

Hoopingarner, R.A. and D.H. Krause (1968) The Mackinac Island fly problem. II. Induced insecticide resistance. Michigan Agricultural Experiment Station Quarterly Bulletin 50:281-284.

Hoy, M.A. and J.J. McKelvey, Jr., eds. (1979) Genetics in Relation to Insect Management. Proceedings of a Conference, March 31 - April 5, 1978, Bellagio, Italy. New York: Rockefeller Foundation.

Huffaker, C.B. and P.S. Messenger, eds. (1976) Theory and Practice of Biological Control. New York: Academic Press, Inc.

Jackson, W.B. (1977) Evaluation of rodent depredations to crops and stored products. EPPO Bulletin 7(2):439-458.

Jackson, W.B. (1979) Second-generation Anticoagulant Rodenticides. Presented at Ninth International Congress of Plant Protection, August 5, 1979, Washington, D.C.

Jackson, W.B., P.J. Spear, and C.G. Wright (1971) Resistance of Norway rats to anticoagulant rodenticides confirmed in the United States. Pest Control 39(9):13-14.

Jackson, W.B., E. Brooks, M. Bowerman, and E. Kaukeinen (1975) Anticoagulant resistance in Norway rats as found in U.S. cities. Pest Control 43(4):12-16; 43(5):14-24.

Jackson, W.B., L. Brown II, and A.D. Ashton (1978) Controlling resistant rats in Chicago. Pest Control 46(8):17, 19-20.

Jackson, W.B. and A.D. Ashton (1979) Present distribution of anticoagulant resistance in the United States. Pages 392-397, 8th Steenbock Symposium: Vitamin K Metabolism and Vitamin-K-dependent Proteins, edited by J.W. Suttie, University of Wisconsin, Madison. Baltimore, Md.: University Park Press.

Johnson, W.T. and H.H. Lyon (1976) Insects That Feed on Trees and Shrubs. Ithaca, N.Y.: Cornell University Press.

Kamble, S.T., R.E. Gold, and E.F. Vitzthum (1978) Nebraska Pesticide Use Survey—Structural. Report No. 1, December. Institute of Agriculture and Natural Resources. Lincoln, Nebr.: University of Nebraska.

Kaukeinen, D. (1979) Experimental rodenticide (Talon) passes lab tests; moving to field trials in pest control industry. Pest Control 47(1):19-21, 46.

Keil, J.E., S.T. Caldwell, and C.B. Loadholt (1977) Pesticide Usage Survey of Agricultural, Governmental, and Industrial Sectors in the United States, 1974. (Abridgement of a report made to EPA, May 31, 1976). Report No. EPA-540/9-78-007, GPO 1979-620-007/3743. Washington, D.C.: U.S. Government Printing Office.

Kenaga, E.E. and C.S. End (1974) Commercial and Experimental Organic Insecticides. Entomological Society of America Special Publication No. 74-1, October. College Park, Md.: Entomological Society of America.

Kennedy, M.K. and R.W. Merritt (1980) Fudge, flies, and dung: Our environment in conflict on Mackinac Island (tentative title). Natural History. (submitted)

Koehler, C.S., W.D. Hamilton, and C.S. Davis (1979) Cypress Tip Moth. University of California Division of Agricultural Sciences Leaflet 2537. Berkeley, Calif: University of California, Cooperative Extension Service.

Kutz, F.W., S. Strassman, and A.R. Yobs (1977) Survey of pesticide residues and their metabolites in humans. Pages 523-539, Pesticide Management and Insecticide Resistance, edited by D.L. Watson and A.W. Brown. New York: Academic Press, Inc.

Kutz, F.W. (1978) Human Monitoring Program. Pesticide Monitoring Semi-Annual Report No. 9, April-September 1978. Washington, D.C.: U.S. Environmental Protection Agency.

Kutz, F.W., R.S. Murphy, and S.C. Strassman (1978) Survey of pesticide residues and their metabolites in urine from the general population. Pages 363-369, Pentachlorophenol, edited by K. Ranga Rao. New York: Plenum Publishers.

Laing, J.E. and J. Hamina (1976) Biological control of insect pests and weeds by imported parasites, predators, and pathogens. Pages 685-743, Theory and Practice of Biological Control, edited by C.B. Huffaker and P.S. Messenger. New York: Academic Press, Inc.

Lande, S.S. (1975) Public attitude toward pesticides, a random survey of pesticide use in Allegheny County, Pa., 1975. Public Health Reports 90(1):25-28.

Latheef, M.A. and R.D. Irwin (1979) The effect of companionate planting on Lepidopteran pests of cabbage. Canadian Entomologist 3:863-864.

Lawless, E.W. and R. von Rumker (1976) A Technology Assessment of Biological Substitutes for Chemical Pesticides. Prepared for National Science Foundation, Contract No. NSF-C-849, MRI Project No. 3879-L. Kansas City, Mo.: Midwest Research Institute.

Legator, M.S., F.J. Kelley, S. Green, and E.J. Oswald (1969) Mutagenic effects of captan. Annals of the New York Academy of Science 160:344-351.

Levenson, H. and G.W. Frankie (In Press) A study of homeowner attitudes and practices toward arthropod pests and pesticides in three U.S. metropolitan areas. Hilgardia.

Lewis, V.R. (1979) A Study of Aggressive Behavior in the Brown-banded Cockroach, *Supella longipalpa* (Fabricius) (*Dictyoptera*, *Blattaria*, *Blattellidae*). M.S. Thesis, University of California, Berkeley, California.

Luck, R.F. and D.L. Dahlsten (1975) Natural decline of a pine needle scale (*Chionaspis pinifoliae* [Fitch]) outbreak at South Lake Tahoe, California following cessation of adult mosquito control with malathion. Ecology 56:893-904.

Maddy, K.T. (1978) Unpublished data, December 1978. Sacramento: California Department of Food and Agriculture.

Makita, T., Y. Hashimoto, and T. Noguchi (1973) Mutagenic, cytogenetic, and teratogenic studies on thiophanate-methyl. Toxicology and Applied Pharmacology 24:206-215.

Metcalf, R.L. (1975a) Insecticides in pest management. Pages 235-273, Introduction to Insect Pest Management, edited by R.L. Metcalf and W.H. Luckmann. New York: John Wiley and Sons, Inc.

Metcalf, R.L. (1975b) Pest management strategies for the control of insects affecting man and domestic animals. Pages 529-564, Introduction to Insect Pest Management, edited by R.L. Metcalf and W.H. Luckmann. New York: John Wiley and Sons, Inc.

Morgan, D.L., G.W. Frankie, and M.J. Gaylor (1978) Potential for developing insect-resistant plant materials for use in urban environments. Pages 267-294, Perspectives in Urban Entomology, edited by G.W. Frankie and C.S. Koehler. New York: Academic Press, Inc.

Munro, J.A. (1963) Biology of the ceanothus stem-gall moth, *Periploca ceanothiella* (Cosens). Journal of Research on Lepidoptera 1(3):183-190.

Munro, J.A. (1965) Occurrence of *Psylla uncatoides* on *Acacia* and *Albizia*, with notes on control. Journal of Economic Entomology 58(6):1171-1172.

National Pest Control Association (1979a) Integrated Pest Management in Food Distribution Warehouses. Technical Release ESPC 072101-10. Vienna, Va.: National Pest Control Association.

National Pest Control Association (1979b) Integrated Pest Management in Retail Food Stores. Technical Release ESPC 072111-18. Vienna, Va.: National Pest Control Association.

National Pest Control Association (1979c) List of Pesticide Chemicals Used by Pest Control Operators by Relative Importance, Type of Compound, Major Pests, and Applications. Draft. Vienna, Va.: National Pest Control Association.

National Pest Control Association (1979d) NPCA Insect Control Committee Insecticide Recommendations. Technical Release ESPC 039999B. Vienna, Va.: National Pest Control Association.

National Pest Control Association (1979e) Rodenticide Use in Food Handling Establishments. Technical Release ESPC 063301A. Vienna, Va.: National Pest Control Association.

National Poison Center Network (1978) Annual Report to the Rohm and Haas Company, First Year 1978. (Unpublished)

National Research Council (1969) Principles of Plant and Animal Pest Control. Volume 3: Insect-Pest Management and Control. Committee on Plant and Animal Pests, Agricultural Board. Washington, D.C.: National Academy of Sciences.

National Restaurant Association (1977) Consumer Reaction Toward Restaurant Practices/Responsibilities. Chicago, Ill.: National Restaurant Association.

National Restaurant Association (1979) Sanitation Operations Manual. Chicago, Ill.: National Restaurant Association.

New York City Department of Health (1977) Annual Report. Rodent Control Program in New York City. Bureau of Pest Control. New York City Department of Health. (Unpublished)

Olkowski, W., D. Pinnock, W. Toney, G. Mosher, W. Neasbitt, R. van den Bosch, and H. Olkowski (1974) An integrated insect control program for street trees. California Agriculture 28(1):3-4.

Olkowski, W., H. Olkowski, R. van den Bosch, and R. Hom (1976) Ecosystem management: A framework for urban pest control. Bioscience 26:384-389.

Olkowski, W., H. Olkowski, T. Drlik, N. Heidler, M. Minter, R. Zuparko, L. Laub, and L. Orthel (1978a) Pest control strategies: Urban integrated pest management. Pages 215-234, Pest Control Strategies, edited by E.H. Smith and D. Pimentel. New York: Academic Press, Inc.

Olkowski, W., H. Olkowski, A.I. Kaplan, and R. van den Bosch (1978b) The potential for biological control in urban areas: Shade tree insect pests. Pages 311-

347, Perspectives in Urban Entomology, edited by G.W. Frankie and C.S. Koehler. New York: Academic Press, Inc.

Olkowski, W., H. Olkowski, and L. Laub (1978c) Urban Pest Control in California: An Assessment and Action Plan. Draft report, Contract No. 9959 for Department of Food and Agriculture, State of California. John Muir Institute for Environmental Studies, Inc., Center for the Integration of Applied Sciences, Berkeley, California.

Painter, R.H. (1951) Insect Resistance in Crop Plants. Lawrence, Kans.: Kansas University Press.

Pederson, J.R. and R.B. Mills (1978) Pest management programs for surplus grain. Pages 369-375, Proceedings of Symposium on Prevention and Control of Insects in Stored-Food Products. Kansas State University, Manhattan, Kansas, April 9-13, 1978. Sponsored by USDA, Kansas State University and Ecological Society of America.

Peoples, S. and K. Maddy (1978) Poisoning due to ingestion of the rodent poison Vacor. Presented at the 8th Meeting of European Poison Control Centers, Netherlands.

Peoples, S. and K. Maddy (1979) Poisoning of man and animals due to ingestion of the rodent poison Vacor. Veterinary and Human Toxicology 21(4):266-268.

Pimentel, D. (1971) Ecological Effects of Pesticides on Nontarget Species. Washington, D.C.: President's Office of Science and Technology.

Pinnock, D.E. (1976) Non-chemical Means of Pest Management in the Highway Landscape. Report No. CA-DOT-MN-4112-76-2. Sacramento: California State Department of Transportation.

Pinnock, D.E., K.S. Hagen, D.V. Cassidy, R.J. Brand, J.E. Milstead, and R.L. Tassan (1978) Integrated pest management in highway landscapes. California Agriculture 32(2):33-34.

Piper, G.L. and G.W. Frankie (1978) Integrated management of urban cockroach populations. Pages 249-266, Perspectives in Urban Entomology, edited by G.W. Frankie and C.S. Koehler. New York: Academic Press, Inc.

Piper, G.L. and G.W. Frankie (1979) Integrated Management of Urban Cockroach Populations. EPA Grant No. R805556010. Washington, D.C.: U.S. Environmental Protection Agency.

Rabb, R.L., R.E. Stinner, and R. van den Bosch (1976) Conservation and augmentation of natural enemies. Pages 233-254, Theory and Practice of Biological Control, edited by C.B. Huffaker and P.S. Messenger. New York, N.Y.: Academic Press, Inc.

Raphael, M.H. (1972) New York City's rat control program. Journal of Milk and Food Technology 33(2):52-58.

Rarig, F.J. (1979) Letter to Dr. Jake MacKenzie, California Department of Food and Agriculture, May 17. (Unpublished)

Reich, G.A., J.H. Davis, and J.E. Davies (1968) Pesticide poisoning in south Florida. An analysis of mortality and morbidity and a comparison of sources of incidence data. Archives of Environmental Health 17:768-775.

Rennison, B.D. (1971) The control of resistant rats and mice. Pages 83-89, Proceedings of Third British Pest Control Conference. London: British Pest Control Association.

Ridgway, R.L. and S.B. Vinson, eds. (1977) Biological Control by Augmentation. New York: Plenum.

Ross, M.H. and D.G. Cochran (1963) Some early effects of ionizing radiation on the German cockroach, *Blattella germanica* (L.). Annals of the Entomological Society of America 56:258-261.

Ross, M.H. and D.G. Cochran (1966) Genetic variability in the German cockroach. I. Additional genetic data and the establishment of tentative linkage groups. Journal of Heredity 57:221-226.

Ross, M.H. and D.G. Cochran (1971) Cytology and genetics of a pronotal-wing trait in the German cockroach. Canadian Journal of Genetics and Cytology 13:522-535.

Ross, M.H. and D.G. Cochran (1973) German cockroach genetics and its possible use in control measures. Patna Journal of Medicine 47:325-337.

Ross, M.H. and D.G. Cochran (1975a) The German cockroach, *Blattella germanica*. Pages 35-62, Handbook of Genetics, Vol. 3, edited by R.C. King. New York and London: Plenum.

Ross, M.H. and D.G. Cochran (1975b) Two new reciprocal translocations in the German cockroach. Cytology and genetics of T(3;12) and T(7;12). Journal of Heredity 66:79-88.

Ross, M.H. and D.G. Cochran (1976) Sterility and lethality in crosses involving two translocation heterozygotes of the German cockroach. Experientia 32:445-447.

Ross, M.H. and D.G. Cochran (1977) Analysis of a double chromosome translocation in the German cockroach. Journal of Heredity 68:231-237.

Ross, M.H. and D.G. Cochran (1979) Properties of a 3-chromosome double translocation heterozygote in the German cockroach. Journal of Heredity 70:259-266.

Sanborn, G.E., J.E. Selhorst, V.P. Calabrese, and J.R. Taylor (1979) Pseudo cerebri and insecticide intoxication. Neurology 29:122-127.

Savage, E.P. (1978) National Household Pesticide Usage Study, 1976-1977, Including Data on Household Poisonings. Draft report, EPA Contract No. 68-01-4663, Colorado State University. (Unpublished)

Seiter, J.P. (1976) The mutagenicity of benzimidazole and benzimidazole derivatives. VX. Cytogenic effects of benzimidazole derivatives in the bone marrow of the mouse and the Chinese hamster. Mutation Research 40:339-348.

Shafik, T.M., H.C. Sullivan, and H.F. Enos (1973a) A multiresidue procedure for halo- and nitrophenols. Measurement of exposure to biodegradable pesticides yielding these compounds as metabolites. Journal of Agricultural and Food Chemistry 21:295-298.

Shafik, T.M., D.E. Bradway, H.F. Enos, and A.R. Yobs (1973b) Human exposure to organophosphorous pesticides: A modified procedure for gas liquid chromatographic analysis of alkyl phosphate metabolites in urine. Journal of Agricultural and Food Chemistry 21:625-629.

Sherer, R.E. (1976) Operation Ben. A Rat Study in Columbus, Ohio. Columbus, Ohio: Ohio Department of Health.

Shternberg, A.I. and A.M. Torchinski (1972) On the interrelationship of genetic toxic, embryotoxic and teratogenic action of chemicals extraneous for the organism and the possibilities for prognosticating their influence on the antenatal period of ontogenesis. Vestnik Akademii Meditsinskikh SSSR 3(27):39-46.

Spencer, E.Y. (1973) Guide to Chemicals Used in Crop Protection. Publication 1093. 6th Edition. Ottawa, Ontario: Agriculture Canada.

Stearns, F. (1978) Urban ecology—opportunity or tar pit? Bulletin of the Ecological Society of America 59:7-9.

Stearns, F.W. and T. Montag, eds. (1974) The Urban Ecosystem, A Holistic Approach. Stroudsburg, Pa.: Dowden, Hutchinson & Ross, Inc.

Tahranainen, J.O. and R.B. Root (1972) The influence of vegetational diversity on the population ecology of a specialized herbivore, *Phyllotreta cruciferae* (Coleoptera: Chrysomelidae). Oecologia (Berlin) 10:321-346.

Thirty-first Illinois Custom Spray Operators Training School (1979) Proceedings. University of Illinois, Urbana.

Todd, F.M. (1909) Eradicating Plague from San Francisco. Report of the Citizen's Health Commission. San Francisco, Calif.: Murdock Press.

U.S. Congress, Senate (1976) Environmental Protection Agency and the Regulation of Pesticides. Staff Report to the Subcommittee on Administrative Practice and Procedure of the Committee on the Judiciary of the U.S. Senate. December 1976. 94th Congress, 2nd Session.

U.S. Consumer Product Safety Commission (1976) National Electronics Injury Surveillance System of Emergency Rooms. Washington, D.C.: U.S. Consumer Product Safety Commission.

U.S. Department of Agriculture (1974) Farmers' Use of Pesticides in 1971—Quantities. Economic Research Service, Agricultural Economic Report No. 252. Washington, D.C.: U.S. Department of Agriculture.

U.S. Department of Agriculture (1978a) Farmers' Use of Pesticides in 1976. Economics, Statistics, and Cooperatives Service. Agricultural Economic Report No. 418. Washington, D.C.: U.S. Department of Agriculture.

U.S. Department of Agriculture (1978b) The Pesticide Review 1977. Agricultural Stabilization and Conservation Service. Washington, D.C.: U.S. Department of Agriculture.

U.S. Department of Health, Education, and Welfare (1969) Report of the Secretary's Commission on Pesticides and Their Relationship to Environmental Health. Parts I and II. Washington, D.C.: U.S. Government Printing Office.

U.S. Department of Health, Education, and Welfare (1977) Human Health and the Environment, Some Research Needs. Report of the Second Task Force for Research Planning in Environmental Health Science. DHEW Publication No. NIH 77-1277. Washington, D.C.: U.S. Government Printing Office.

U.S. Environmental Protection Agency (1974) Strategy of the Environmental Protection Agency for Controlling the Adverse Effects of Pesticides. Office of Pesticide Programs, Office of Water and Hazardous Materials. Washington, D.C.: U.S. Environmental Protection Agency.

U.S. Environmental Protection Agency (1976) National Study of Hospital Admitted Pesticide Poisonings. Epidemiologic Studies Program, Human Effects Monitoring Branch, Technical Services Division, Office of Pesticide Programs. Washington, D.C.: U.S. Environmental Protection Agency.

U.S. Environmental Protection Agency (1978) U.S. Rodenticides and Their Regulatory Status in Commensal Rodent Control. Office of Pesticide Programs, Registration Division. Washington, D.C.: U.S. Environmental Protection Agency.

U.S. Environmental Protection Agency (1979a) Pesticide product information on microfiche (approximately 370,000 pages). Office of Pesticide Programs. Washington, D.C.: U.S. Environmental Protection Agency.

U.S. Environmental Protection Agency (1979b) Selected pesticide product information computer-printed for the NRC Committee on Urban Pest Management. (Approximately 6,500 pages.) Office of Pesticide Programs. Washington, D.C.: U.S. Environmental Protection Agency.

U.S. Environmental Protection Agency (1979c) List of principle (sic) urban pesticides and their uses (March 9, 1979). Office of Pesticide Programs, Registration Division. Washington, D.C.: U.S. Environmental Protection Agency.

U.S. Environmental Protection Agency (1979d) Pesticide Industry Sales and Usage, 1979 Market Estimates. (7 tables.) Office of Pesticide Programs. Economic Analysis Branch. Washington, D.C.: U.S. Environmental Protection Agency.

U.S. Public Health Service (1979) Urban rat control—United States, April-June 1979. Center for Disease Control. Morbidity and Mortality Weekly Report 28(42):505-507.

van den Bosch, R. and A.D. Telford (1964) Environmental modification and biological control. Pages 459-488, Biological Control of Insect Pests and Weeds, edited by P. DeBach. London: Chapman and Hall.

von Rumker, R., R.M. Matter, D.P. Clement, and F.K. Erickson (1972) The Use of Pesticides in Suburban Homes and Gardens and Their Impact on the Aquatic Environment. Pesticides Study Series 2. Office of Water Programs, U.S. Environmental Protection Agency. Washington, D.C.: U.S. Government Printing Office.

von Rumker, R., E.W. Lawless, A.F. Meiners, K.A. Lawrence, G.L. Kelso, and F. Horay (1974) Production, Distribution, Use and Environmental Impact Potential of Selected Pesticides. Report No. EPA-540/1-74-001. Office of Water and Hazardous Materials, Office of Pesticide Programs. Washington, D.C.: U.S. Environmental Protection Agency.

Voogd, C.E., J. Jacobs, and J. Vander Stal (1972) On the mutagenic action of dichlorvos. Mutation Research 16:413-416.

Walcott, R.M. and B.W. Vincent (1975) Rats, Fires, and Inner-City Solid Waste Storage Practices. Report No. EPA-530/SW/150. Cincinnati, Ohio: U.S. Environmental Protection Agency.

Weidhaas, J.A., Jr. (1976) Is host plant resistance a practical goal for control of shade-tree insects? Pages 127-133, Better Trees for Metropolitan Landscapes, edited by F.S. Santamour, Jr., H.D. Gerhold, and S. Little. USDA Forest Service General Technical Report NE-22. Washington, D.C.: U.S. Government Printing Office.

Whorton, D., R.M. Krauss, S. Marshall, and T.H. Milby (1977) Infertility in male pesticide workers. Lancet 2(8051):1259-1261.

Williams, M.L., C.H. Ray, and I.E. Daniels (1977) The euonymus scale in Alabama. Highlights of Agricultural Research, Auburn University Agricultural Experiment Station Bulletin 24(2):15.

World Health Organization (1976) Resistance of Vectors and Reservoirs of Disease to Pesticides. Technical Report Series 585. Geneva: World Health Organization.

Yobs, A.R. (1971) The national human monitoring program for pesticides. Pesticide Monitoring Journal 5(1):44-46.

4

Economics of Urban Pest Management

This chapter[1] examines some theoretical aspects of the social efficiency of current urban pest management practices by investigating the possibility that society either forgoes important benefits or pays excessively high costs under the current configuration of urban pest management practices and institutional arrangements. There are strong reasons for believing that the private market alone cannot produce socially efficient solutions and that to the extent that society relies on the private market it incurs unknown but probably substantial costs in pursuit of urban pest control.

A CONCEPTUAL FRAMEWORK

This analysis of the social efficiency of urban pest management focuses on two related questions. First, is the current level of resource allocation appropriate, or are there benefits to be gained from increasing or decreasing the level of activity devoted to urban pest control? Second, are current practices cost-effective or are there better ways of managing urban pests? The first question requires an analysis of the relative costs and benefits associated with various levels of control. The second question asks whether the costs of control could be lower, given the existence of alternative methods.

PRIVATE INTEREST VERSUS SOCIAL EFFICIENCY

Greater social efficiency in pest management would increase the net benefits of pest control. Several factors account for the failure of the private market mechanism to produce a more socially desirable result: (a) lack of appropriate economic incentives; (b) inadequate information; (c) highly unequal distribution of income and the existence of extreme poverty; and (d) inappropriate or inefficient government regulations and other institutional barriers.

Without external intervention there is no mechanism to force private pest control firms or individuals to take into account the impact of their actions on the welfare of others. Thus, resources may be allocated inefficiently from a number of perspectives. If one person chooses to let his or her property deteriorate and become infested with rats and other pests, for example, the effects of that decision will be felt throughout the neighborhood. The individual property owner, however, does not calculate the cost of maintaining the property in terms of the costs that deterioration and pest infestation impose on society, but only in terms of the effect of the cost on his or her own welfare. As a result, what may be rational from a private profit-maximizing perspective may be harmful from the perspective of society. In economic terms, the relationship of the property owner to the neighborhood described here is called an adverse "externality."

Another example of an adverse externality is the impact of toxic chemicals on the environment and on human health. There is little doubt that some chemicals used in urban pest management have important adverse side effects. Since such toxic chemicals also offer cheap and effective control of pests, however, it is very likely that the private market will encourage their excessive use, thus causing a divergence between private and social efficiency. That is, the individual user will tend to underestimate the full costs of using toxic chemicals and seek to maximize private profit regardless of the substantial social costs that may accrue. If all of the social costs were imposed on the user, individuals would have a substantial incentive to use less of the chemicals.

Even if this degree of social efficiency could be brought about through public intervention in the market, it is still possible that some people would be worse off. Social efficiency means only that total benefits are at their greatest level. This situation can result when the costs imposed on society are more than offset by benefits to the individual user. In other words, maximum social efficiency does not necessarily mean that the distribution of costs and benefits is equitable. This is an important point, for as some point out, if one person is made worse off, the society may be worse off, no matter how large the benefits produced. If a more strict definition of social efficiency employing distributional considerations is used, the conclusion that the private market process creates incentives to overuse toxic chemicals is strengthened.

Pest resistance is another example of an adverse externality. Individual users or pest control operators are unlikely to take into account the consequences of their current behavior on future pest control. A private applicator who voluntarily reduced the use of profit-maximizing chemicals would find current costs increasing and profits decreasing and yet, if none of the competing pest control operators curtailed their use of chemicals, pests would become resistant to chemical pesticides anyway. Thus over time, pest resistance would be likely to occur no matter what course of action a single individual takes, and not to take advantage of currently available profits would make no economic sense. Here, again, it is reasonable to believe that the private market will not produce the long-term socially desirable allocation of resources.

It might be argued that if private actions that relate to pest control impose large social costs, those on whom the costs are imposed will have an incentive to organize to force heavy users of chemicals (or landlords with infested buildings) to take more appropriate action. Such collective action is unlikely, however. No single individual is likely to be affected strongly enough to warrant acceptance of the cost of organizing the rest of the affected community, which may be very high. Moreover, there are economic incentives that work against collective action. Theoretically, at least, each individual realizes that collective home repair or yard cleanup or restricted use of chemicals will benefit him or her even if he or she does nothing. Therefore, the tendency is for everyone to let everyone else solve the problem. This is the classic "free rider" problem, and contributes to the condition of social costs in urban pest management.

In addition to adverse externalities, social inefficiency may also result from inadequate information. An individual user who is not aware of a relationship between chemical exposure and human health, for instance, will not take the necessary precautions. Poor information may also cause waste. Chemical control of insects, weeds, and other pests, for example, is frequently a complex process requiring expertise not usually possessed by the individual householder. The wrong chemical or the correct one applied at the wrong time or in the wrong amount may lead to large social as well as private costs, with no compensating benefit. Finally, lack of information may lead some persons to underallocate pest control resources. The individual who ignores, for example, the possible damage by termites may experience much larger costs over the long run than the person who regularly monitors the home for termites and takes appropriate action if they are found. Conversely, some pest control activities would likely be discontinued if, for example, the homeowner were made aware that the pest posed no health risk or long-term damage to ornamental plants.

There are other persons who will not be able to take appropriate action to control pests because they must spend all their resources on more immediate needs. In such cases the social costs of uncontrolled pests are likely to be reflected in additional human health problems, in additional demands on public welfare services, and in general economic decline of the community. Since poverty is usually associated with living conditions conducive to rapid pest population growth, and since the poor often lack either the resources or the incentives to effectively control pests, it is reasonable to hypothesize that pest problems will be greatest among the poor.

CHOICE OF TECHNOLOGY

In theory, there is a wide range of urban pest management strategies. Thus, it is worth asking whether there are incentives for urban users to minimize the use of chemicals when other choices are available. For without strong incentives, alternative approaches will not be adopted unless they are at least as profitable from a private perspective as the traditional chemical approach. Even when incentives exist, however, there may be other reasons why alternate pest management technologies will not be employed (or even sufficiently developed) (Olkowski and Olkowski 1978).

Approaches that minimize chemical use are described in the growing literature on integrated pest management (IPM). Four components of IPM are relevant to this analysis. First, IPM implies the existence of a system for regular monitoring of a pest population and its natural enemies. Monitoring also includes keeping track of environmental conditions that affect the size of pest populations, such as temperature, humidity, sanitation, and the like.

A second component of IPM is determination of the relationship between the size of the pest population and the degree of damage—whether aesthetic, health, or economic—that it causes. From this basic relationship an economic damage function can be derived (at least in theory) by quantifying the various damages and attaching economic values to them. The extreme difficulty of actually deriving damage functions for pest problems is clearly illustrated by the inability of investigators, to date, to develop sufficient information to obtain such functions. Thus, while in theory the derivation of damage functions could prove useful in the analysis of pest management problems, there has yet to be a demonstrated and reliable determination of damage functions.

A third component of IPM is identification of all the available control methods, including physical, cultural, biological, and chemical methods, and their relative costs. Out of this analysis it is possible to develop a strategy combining several control options that are appropriate to achieve the levels of control desired. A comparison of costs with the benefits to be derived from differing levels of pest population size suggests what degree of control is profitable and when, where, and how it can be undertaken.

Finally, IPM requires an evaluation process in which the effects of a given strategy are fed back to the decision maker for evaluation and possible revision.

The advantages of IPM over the conventional pesticide approach are that it provides stable long-term control over pest populations, minimizes the use of chemicals and hence their impact on human health and the environment, and may offer the lowest-cost tactics consistent with long-run stability. It might be added that since IPM strategies rely on chemicals to some extent, they do not necessarily produce control levels consistent with social efficiency as defined above. Nevertheless, IPM seems likely to produce more socially efficient results than sole reliance on chemical pesticides.

As noted in Chapter 3, IPM strategies have been employed far more in agriculture than in urban areas, but even in agriculture acceptance has been slow (Willey 1978). IPM strategies are much less well-defined for urban areas, and there are many reasons to believe that IPM acceptance among urban users will be even slower than in agriculture. The reasons are related to the relative profitability of conventional and IPM strategies, to the more stringent monitoring and information requirements of IPM strategies, to institutional and regulatory barriers, and to a variety of other noneconomic factors.

Although IPM strategies reduce pesticide use, they require skilled human labor to monitor pest populations and to design and implement successful strategies. In agriculture, much of the research on IPM is being done in the public sector and is thus subsidized. Monitoring or scouting techniques are being developed to allow individual farmers to relay information about pest problems to a centralized facility which then advises them of the appropriate actions to take. In other words, large-scale application of IPM combined with public subsidies has made it possible to substitute skilled labor for chemicals and to reduce the high costs of this substitution. In these circumstances IPM programs can compete with conventional programs, and farmers who adopt IPM are not compelled to give up profits.

In the urban environment, on the other hand, there are few comparable settings for the development of large-scale IPM systems. Only if IPM programs offer lower costs will local agencies have sufficient incentive to make the initial investment. As governmental budgets are increasingly squeezed, the choice of technology is likely to be determined by budgetary factors that influence agencies to adopt less expensive methods of pest control. Skilled workers are a fixed cost that cannot be easily manipulated, while chemical controls can be used without additional skilled workers and thus offer a government agency greater budgetary flexibility. Thus, even when the two control methods are equally effective, the chemical approach may be the more rational choice from the perspective of the government decision maker.

For the majority of urban uses, IPM strategies may be more costly in the short term and less costly in the long term, than conventional approaches. Information about IPM is not yet readily available, and the typical user concerned with home and garden pest problems must invest considerable time in finding information about IPM. Information on chemical controls, on the other hand, is readily available from nurseries, hardware stores, and other places where pesticides are sold. Since obtaining this information is cheaper, it helps to make chemical controls cheaper. Chemical controls are also "subsidized" by chemical firms through advertising, funding demonstrations by research groups, and by other means.

IPM strategies also involve continuous monitoring of the pest population, and here again expertise is needed. The typical backyard gardener will find it easier to read the label on a pesticide container than to identify an insect, the stages of its life cycle, the size of its population, and the likelihood that it will produce undesirable damage, and then to determine the various types of actions needed for successful IPM. Although attempts should be made to "package" IPM strategies for small users, the diversity of urban environments in any given region and the diversity among regions are likely to make widespread dissemination of IPM controls for home and garden pests difficult to achieve. As long as chemical controls offer cheap, convenient, and effective short-term control, they are likely to be preferred.

It might be argued that some commercial pest control businesses are large enough to invest in the information needed to develop successful IPM strategies. Here again, however, economic logic dictates otherwise. First, unless chemicals become much more expensive, there may be no incentive to use additional skilled labor unless pest resistance significantly lowers the effectiveness of existing chemicals. Chemicals offer the pest control operator (PCO) higher profits and effective short-term control, which is likely to be more desirable to the PCO than long-term control, because it offers the opportunity to develop a large number of repeat customers. Although routine preventive spraying implies substantial use of chemicals, it may be cheaper than developing a monitoring system and spraying only when there is evidence that the pest population has gotten large enough to warrant action. Moreover, given the facts that many customers believe that effective control means total eradication of the pest population, and that many local codes and ordinances specify zero-level pest populations in food-handling establishments, the PCO may believe that complete eradication is necessary if the business is to survive. A pest population of zero is the antithesis of the IPM approach, which seeks to manage pest populations rather than to eradicate them. Thus, customers' attitudes or the operator's perception of those attitudes may force the PCO to utilize only the chemical approach.

Another barrier to IPM that exists in all urban settings is the fragmentation of control over the environment by any one firm or individual. To be effective, IPM requires control over a sufficient portion of the natural environment. This is particularly true when biological controls are used. If some homeowners use insecticides, the chemicals may drift into neighboring yards and disrupt natural control processes that might otherwise be effective. If such behavior cannot be changed, IPM will offer less protection and hence will be used less. If neighborhoods develop areawide controls, the cost of the investment can spread over the entire area, thus making the approach more cost-effective. Given the diversity of interests in neighborhoods, however, there are great problems involved in achieving collective action.

Another problem is that governmental regulations may be biased against IPM. Local codes and ordinances specifying zero-level pest populations as the maximum acceptable infestation level would, in the absence of regulations on maximum permissable pesticide body burden for residents, encourage heavy application of chemicals. At the same time, biological controls that permit a certain number of pests and their natural enemies do not satisfy such regulations. Furthermore, such regulations do not permit a determination of the economic damage levels below which it simply does not pay to control the pest population. When regulations require control below the economic damage level, they guarantee economically inefficient allocation of pest control resources.

Finally, IPM strategies are not consistent with prevailing public attitudes. Widespread insect and rodent phobias among the public lead to a demand for zero-level pest populations. Such absolute control is unlikely to occur with IPM. These attitudes may change as the consequences of exposure to toxic chemicals become more widely understood, but for the present they play a significant role in preventing the development of interest in alternative urban pest management approaches.

DISTRIBUTIONAL IMPLICATIONS OF URBAN PEST MANAGEMENT ACTIVITIES

Since it is reasonable to assume that many pest problems are relatively greater in low-income areas than in higher-income neighborhoods, it follows that pest-related health costs and other costs are also likely to be greater in poorer neighborhoods. This would seem to be especially true because of the inability of persons living in poorer areas to control their environment through private pest control measures. It may also follow that any pest control activities that are undertaken in poorer areas will be concentrated within households and focus for the most part on pests that directly threaten residents—rodents, cockroaches, and ants, rather than termites or outdoor pests. Low-income groups may also be more prone to use pest control methods that require the least expenditure. This often means using products purchased at retail stores and directly applied by household members, which suggests that pesticides are frequently misused and overused.

Higher-income groups, on the other hand, are more likely to live in well-maintained neighborhoods with lower overall levels of human health-related pest infestation. Exposure to pest-related problems is likely to be lower as well. Moreover, higher-income groups are likely to control pest populations more effectively because of important social incentives, and if pest problems should occur there are fewer disincentives to collective action. Higher-income groups are also likely to place greater emphasis on controlling pests that threaten physical damage, since they usually own their homes and would experience economic loss if their property deteriorates.

It should be pointed out, however, that even though there may be a lower overall level of pests in higher-income neighborhoods, it is possible that the total amount of pesticides applied there exceeds application in low-income neighborhoods.

Whether greater use of pesticides in higher-income areas results in greater human exposure to pesticides is not clear, however. If higher-income groups rely more on PCOs, and if these firms are more careful in applying pesticides or achieve greater control using smaller amounts than would the inexperienced layman, it is possible that even though more pesticides are used, human exposure is lower. If, as some studies indicate, income and education are highly correlated, higher-income groups are also more likely to be more aware of the dangers of toxic chemicals and to take greater precautions in applying them (Levenson and Frankie, in press).

Thus, it is highly possible that low-income groups suffer greater levels of pests, pest-related health and nuisance costs, pesticide exposure, and the problems caused by such exposure than do higher-income groups. Empirical verification of this contention would be useful, but data that would allow careful analysis do not exist. There are clues in the survey work of Levenson and Frankie (in press) that suggest some of the distributional implications discussed above, but these surveys deal with attitudes and not with actual behavior. There are also indications that residues of certain chemicals (such as DDT and lindane) are greater in the tissues of blacks than in whites throughout the United States (Davies et al. 1972) and that residues in urban soil are at least as high as those in agricultural land (Carey et al. 1976). These findings may be explained by other factors, such as diet, but they also suggest higher pesticide exposures within low-income households (Davies et al. 1972).

IMPLICATIONS FOR PUBLIC INTERVENTION

Urban pest management activities may warrant public regulation for two reasons. First, the private market may not adequately control pests that impose high costs on certain people and places within urban areas, and public intervention may be warranted to ensure that the potential benefits of urban pest control are fully realized. Second, because pesticides may impose a variety of social costs that are not taken into account by the firms and individuals who use them, there is a strong possibility that pesticides are overused (that is, their marginal costs exceed their benefits) and that other pest control techniques are not developed or adopted.

ESTIMATING COSTS AND BENEFITS OF PEST MANAGEMENT STRATEGIES

Implicit in the discussion to this point has been the assumption that the consequences (costs and benefits) of various pest control strategies can be measured and compared. This assumption may be valid in analyzing the private costs and benefits that individuals and firms use in their cost calculations, but questions about social costs and benefits cannot be resolved through reference to the "facts." There simply are no facts on which to base strong conclusions.

METHODOLOGICAL ISSUES

The traditional analytical approach to comparing costs and benefits is based on the concept of incremental, or marginal, costs and benefits. That is, the maximum benefits to be derived from any strategy occur when the last increment of control yields benefits equal to the costs. If a control effort produces marginal benefits greater than marginal costs, too little effort is being made.

Measuring costs and benefits requires establishing the physical relationship at each level of control and the consequences of that particular level (such as reduced property damage or greater health risk). When pest management strategies are compared, similar analyses must be made for each strategy.

An honest appraisal of the likelihood of successfully completing such an analysis—given the severe constraints imposed by limited understanding of such important relationships as those between human health and pesticide exposure, and by the lack of such essential information as who uses pesticides, for what purposes, and in what quantities—must conclude that all that is possible now is to lay the foundation for later analysis. But if it is impossible to undertake a detailed marginal analysis, it may still be possible to draw suggestive conclusions on the basis of incomplete information.

If, for example, the social costs of chemical controls can be shown to be very large or the benefits relatively small, or if there are effective and reasonably competitive alternative methods of control that minimize the use of pesticides, change in current urban pest management practices may be justified. If, on the other hand, the benefits of chemical controls can be shown to be very large, and if there are no viable alternatives at reasonable costs, the case for continuing current practices may be justified, even if there are associated social costs. Since the primary issue here is not whether there is a need for fine-tuning of the system but whether there is a need for fundamental restructuring of the system, a general analysis may provide sufficient guidance.

THE MEASUREMENT OF SOCIAL COSTS

The most easily identified social costs of pesticides are adverse effects on human and animal health, and the effects of pest resistance on health or pest management expenditures. In order to quantify health costs, one would need to know the relationship between levels of exposure to a given chemical and the probable symptomatic health effects. Obtaining this knowledge would be a major undertaking, even if basic relationships between exposure and health were well understood. At present, however, there are no comprehensive epidemiological studies of the relationship in urban areas between pesticide exposure levels and their effects on health.

As noted in Chapter 3, studies of pesticide-induced illness and disease in particular populations do exist, although most of these have involved persons in agricultural or occupational settings.

It might be thought that urban populations are exposed to much lower levels of pesticides than farm workers or other occupational groups, such as pest control operators. Although this may be true, there is also considerable evidence that pesticides are very prevalent in urban areas. Thus, everyone in urban areas may be exposed to varying degrees. Von Rumker et al. (1972) indicate that on a per-acre basis urban pesticide applications, especially of fungicides and insecticides, are larger than in agriculture. Nonetheless, the difficulty of determining with any precision the health effects of urban pesticide use and of assigning dollar values to these effects makes it impossible to estimate the health costs. Although various scenarios have been developed to assist in environmental decision making (e.g., Lave and Seskin 1977), the area of the estimation of the value of human life and the value of other health and aesthetic effects is clearly one of great controversy.

Another social cost arises from insect and rodent resistance to pesticides. The net effect of pest resistance is that greater quantities of existing chemicals, or the creation of new ones, are required for effective control. Hence, the cost of chemical controls is increased. If pest resistance completely eliminates the possibility of effective chemical control, the cost to society is increased still more by the consequent health problems and property damage. Thus far, however, except for a few agricultural pests, the long-term costs of pest resistance have not been carefully studied (Regev et al. 1977). The cost of current pesticide applications on future levels of pest resistance and pest control are therefore not known. These costs may be high, however, if control over important pest-borne diseases is lost, as may be the case with the loss of control of some species of mosquitoes. However, if effective nonchemical controls can be implemented in cases where resistance has developed, then the overall costs of this phenomenon may not be crucial.

THE MEASUREMENT OF PESTICIDE BENEFITS

A second component of the debate over conventional pest management practices concerns the magnitude of the benefits that pesticides provide. These benefits are generally classified as economic, health, and quality-of-life or aesthetic benefits.

Economic benefits generally refer to the class of impacts that describe the protection of property. Control of termites and other wood-destroying pests provides such benefits, as does control of insects that destroy carpets, cloth, crops, and trees. Protection of golf courses, parks, and industrial facilities against pests also yields economic benefits. Another class of economic benefits results from efforts undertaken to protect a business by meeting an environmental or health code or regulation (such as is required of restaurants) or to maintain an environment conducive to good business (such as keeping objectionable insects out of hotel rooms).

Economic benefits are routinely calculated for agricultural pesticides, but determination of the economic benefits of pesticides in urban environments is made difficult by a paucity of data. It must also be recognized that if pesticide use is restricted, other practices are likely to be implemented to prevent economic loss. Greater emphasis might be placed on determining the level of pest damage before applying pesticides with the resultant elimination of routine spraying. The alternative practices might lead to greater damage or to higher costs of protection while reducing the cost of chemical applications. To assess the real benefit of current urban pest management practices on property values, the analysis must consider the effectiveness and costs of alternatives for each level of control desired.

When information needed to calculate the benefits of protecting property with pesticides is missing, a common approach is to determine the cost of chemical controls and use this as a surrogate for benefits. The assumption is that if individuals are willing to spend a certain sum to protect themselves and their property, they must derive at least that much economic benefit. The surrogate must be used with care, however, because it is not a measure of net benefits but of gross benefits, since no consideration is given to what might have been paid for alternative methods of control or to the effects of these controls.

The same kinds of problems encountered in estimating health costs of pesticide exposure are encountered in estimating health benefits. The literature on the health benefits accruing from pesticide use emphasizes the health hazards associated with various pests and not the effects of varying levels or types of control on these health hazards. As in the case of property damage it is necessary to know whether other, perhaps more costly, methods of control would be employed in the absence of pesticides. No such information exists, although it might be possible to conduct epidemiological studies of differences in the incidence of certain diseases in order to estimate the probable impact of measures taken to control particular diseases.

Surveys (see, for example, von Rumker et al. 1972) demonstrate that most of the pesticides used in urban areas is for the purpose of protecting property or for aesthetic benefits rather than for disease control. One implication of this is that overall pesticide use in urban areas could be reduced substantially without threatening public health. Even if it is assumed that the pesticides used to prevent diseases have large benefits, it need not follow that these benefits must be sacrificed in order to reduce pesticide use.

The intangible aesthetic benefits of pesticides must also be considered. In this case the value of pest control using pesticides is largely a matter of personal preference. The usual practice for evaluating aesthetic benefits is to determine the costs to the beneficiary of controlling a given pest. The assumption is that a person willing to pay for control must receive at least as much benefit. As noted earlier, the costs of control can provide a surrogate estimate for the perceived benefit of pest control. But the benefit of any particular form of pest control, such as pesticide use, is only the net increase in benefits relative to the next best control strategy. If, for example, there is another method of growing roses that is as effective as, but more costly than, using pesticides, the incremental value of the pesticide is the difference between the costs of the two strategies.

Because aesthetic benefits cannot be measured objectively, the question arises as to the weight they should be accorded in an overall evaluation. Neither this Committee, nor any other, can answer that question definitively.

FINAL OBSERVATIONS ON COSTS AND BENEFITS

The preceding observations have a common theme: Given the information currently available, cost/benefit analyses of pesticide-based urban pest control programs cannot be made. At best, the information that exists at present can provide only general guidance about future directions for urban pest management.

The discussion has shown that there are strong reasons for believing that current practices impose social costs that, while difficult to quantify, are likely to be substantial. This, in turn, requires an attempt to define a course of action that reduces pesticide use while still providing effective pest management that does not impose large health or economic burdens on urban residents. Integrated pest management provides for such a course of action, although much more information is needed before it can be conclusively demonstrated that IPM is an effective and efficient urban pest management tool.

Some generalizations may assist in defining a direction for future research and policy. First, the data and studies reviewed in Chapter 3 (see section on Survey of Chemical Controls) indicate that most urban uses of pesticides, excluding rodenticides, are not focused on public health pests to reduce the risk of human disease but on pests that attack plants and structures; and that most of the pesticides applied in urban areas are either rodenticides or insecticides. Rodenticides comprise less than 10 percent of all pesticides (see Table 3.2), and it is assumed that their use is related entirely to public health. Although the actual amounts of pesticides used for public health and other purposes are not known, Olkowski et al. (1978) cite one study in which 60 percent of all homeowner-purchased pesticides was reported to be used in gardens. If the proportion of homeowner-purchased insecticides used in gardens is the same for all homeowner-purchased pesticides, then a major portion of insecticides applied in urban areas is used to protect plants. The amount of insecticides used to control public health pests (e.g., fleas, lice, bedbugs, ticks, mites, mosquitoes) accounts for only a small portion of all insecticides used in urban areas— probably less than 20 percent. For example, von Rumker et al. (1972) found that only 1 percent of all insecticides used in three suburban areas was used in mosquito abatement, the most important of public health concerns, except, perhaps, for rodent control. Combining rodenticides (less than 10 percent) and insecticides (less than 5 percent) yields a relatively small percentage of pesticides used in urban areas for the protection of public health—probably not more than 15 percent.

An interesting implication of this small percentage is that most pesticides are introduced into the urban environment either to protect property or to provide aesthetic benefits. If it is presumed that large benefits are derived from chemicals used to prevent disease, it need not follow that these benefits must be sacrificed in order to reduce the overall level of pesticide use in urban areas, since substantial reductions in pesticide use may be achieved without threatening public health programs. Thus, one important avenue for exploration is the nature of the benefits and costs of those chemicals used primarily to maintain property values or aesthetic qualities.

A second generalization is that if large amounts of chemicals are being introduced for aesthetic purposes, then placing restrictions on the use of such chemicals would reduce the overall amount of pesticide use without impairing the benefits to be derived from property protection or reduced risk of human disease. This implication would be strengthened if more was known about the social costs—toxic effects and pest resistance—associated with the various kinds of pesticides and their uses.

Finally, if effective IPM strategies were available for the control of important urban pests without imposing large additional costs on users, increased restrictions on pesticides used for purposes other than protection of human health would likely increase the social efficiency of resource allocation in urban pest management. For example, use of herbicides saves costs of weed control by replacing manual labor. If labor is unemployed, or if the additional costs of labor are only slightly larger than the cost of chemicals, further restrictions on herbicide use would lead to little social loss of welfare.

A RESEARCH AGENDA FOR THE ANALYSIS OF
URBAN PEST MANAGEMENT STRATEGIES

As noted earlier, there are good theoretical reasons for believing that
current methods of urban pest control are neither socially efficient nor
equitably distributed. Before we can make reasonable judgments on
whether and how present practices should be changed, however, we must
know the answers to some important questions.

SOCIAL EFFICIENCY

In examining the issue of social efficiency, we need to know much more
about the benefits of urban pest control. Estimates of crop loss to certain
pests and the effectiveness of both chemical and nonchemical controls are
available for agriculture, but there is no comparable information on the
reduction of pest-related health problems and economic damage or the
improvement in the quality of life related to either chemical or nonchemical control strategies in urban areas. What are the short- and long-term
health implications of excessive reliance on chemicals to control urban
pests? What are the possible pesticide-induced acute and chronic diseases,
and can the incidence of these diseases be related to changing levels of pest
populations or to different levels of pesticide exposure?

The preceding discussion also raised important questions about the
measurement of aesthetic benefits, and this, in turn, leads to questions
about personal preferences. Are demands for pest-free homes or gardens
related to psychological phobias, to fears of social stigma, to rational or
irrational beliefs about pest-induced disease? To what extent are these
preferences shaped by advertising and other kinds of information about the
negative effects of pest-infested homes, for example, and the benefits of
chemical pesticides? How can the value of various aesthetic benefits such
as weed-free lawns, unblemished roses, or snail-free gardens be taken into
account?

The discussion also identified a set of problems associated with the
development, dissemination, and implementation of integrated pest management strategies that would rely less on chemicals. Does the private
market have any incentive to develop such strategies? Are there obstacles,
either institutional or economic, that prevent alternatives from being
developed? Are the consequences of pest resistance and the rising cost of
oil (and hence chemicals) taken into account in the market place?

A final set of questions relating to social efficiency concerns the role of
the public sector in allocating pest control resources. What has been the

contribution of publicly supported research to the development of alternative kinds of pest controls for the urban environment? Have chemical controls been more heavily subsidized than IPM strategies? What are the effects of public regulations, including health and housing and building codes, on the use of pesticides and the adoption of IPM strategies? Finally, do current EPA procedures for regulating pesticides pose any impediments to managing urban pests or restrict the allocation of resources for controlling urban pests?

SOCIAL EQUITY

Although the association of poverty with pests and pest-related disease and property damage is well established, there has been little analysis of the importance of this association for the individual or for society. To what extent are social benefits to be derived from increasing pest control among low-income groups? What effects would increased pest control have on the public treasury in terms of, say, the demand for public health services? Would increased pest control have a measurable impact on worker productivity or the learning ability of children? Can public investments in pest control programs significantly improve the quality of life in low-income urban areas, or are pest problems simply a symptom of larger problems that can only be ameliorated by raising incomes? To what extent can the public sector help overcome the barriers to collective action that may underlie pest problems in low-income areas? Finally, what is the nature of pesticide exposure in low-income areas? Is it greater than in higher-income areas? Are there particular pesticide-related diseases or acute health hazards that are more prevalent among low-income groups, and how might these health problems relate to the overall problem of poverty?

None of these questions can be answered without a major commitment of funds for research.

NOTE

1. This chapter is based on the working paper, "The Economics of Urban Pest Management" by LeVeen and Flint (see Appendix B).

REFERENCES

Carey, A., G. Wiersma, and H. Tai (1976) Pesticide residues in urban soils from 14 U.S. cities, 1970. Pesticide Monitoring Journal 10:54-60.

Davies, J., W. Edmundson, A. Raffonelli, J. Cassady, and C. Morgade (1972) The role of social class in human DDT pollution. American Journal of Public Health 96:334-341.

Lave, L.B. and E.P. Seskin (1977) Air Pollution and Human Health. Baltimore: Johns Hopkins University Press.

Levenson, H. and G.W. Frankie (In Press) A study of homeowner attitudes and practices toward arthropod pests and pesticides in three U.S. metropolitan areas. Hilgardia.

Olkowski, H. and W. Olkowski (1978) Some advantages of urban integrated pest management programs and barriers to their adoption. *In* Proceedings of the Integrated Pest Management Seminar presented by the University of California Cooperative Extension Service in cooperation with the Los Angeles Commissioner's Office and the Southern California Turfgrass Council, Forest Lawn, Los Angeles, March 1978.

Olkowski, W., H. Olkowski, T. Drlik, N. Heidler, M. Minter, R. Zuparko, L. Laub, and L. Orthel (1978) Pest control strategies: Urban integrated pest management. Pages 215-234, Pest Control Strategies, edited by E.H. Smith and D. Pimentel. New York: Academic Press, Inc.

Regev, U., H. Shalit, and A. Gutierrez (1977) Economic conflicts in plant protection: The problems of pesticide resistance; theory and application to the Egyptian alfalfa weevil. Pages 281-299, Proceedings of a Conference on Pest Management, October 25-29, 1976, edited by G.A. Norton and C.S. Holling. Luxenburg, Austria: International Institute for Applied Systems Analysis.

von Rumker, R., R.M. Matter, D.P. Clement, and F.K. Erickson (1972) The Use of Pesticides in Suburban Homes and Gardens and Their Impact on the Aquatic Environment. Pesticides Study Series 2. Office of Water Programs, U.S. Environmental Protection Agency. Washington, D.C.: U.S. Government Printing Office.

Willey, W.R.Z. (1978) Barriers to the diffusion of IPM programs in commercial agriculture. Pages 285-308, Pest Control Strategies, edited by E.H. Smith and D. Pimentel. New York: Academic Press, Inc.

5

Urban Pest Management Decision Making

RESPONSIBILITY FOR URBAN PEST MANAGEMENT

THE FEDERAL RESPONSIBILITY

This section of the report discusses the role of various federal agencies in urban pest management. It should be made clear at the outset, however, that no federal agency has any legal authority to make policy for urban pest management. Indeed, the most noteworthy comment that can be made on federal involvement in urban pest management is that Congress and the Executive Branch do not have a policy on the subject. Although there is great concern over the management of pesticides and some concern over the management of pests in the agricultural setting, there is literally not a single mention of urban pest management in the federal statutes, with the possible exception of rodent control. To be sure, there is a fair amount of law that has an impact on urban pest management, such as authorization of research on pests, and there are also many federal agencies whose activities have an indirect impact on the management of pests in urban areas. But the existing federal legislation that affects urban pest management reflects no consistent policy. Federal Housing Administration (FHA) concerns about wood-destroying insects and General Services Administration (GSA) contracts for pest control in federal buildings may affect urban pests, but they do not constitute a policy.

In the course of preparing this report, the Committee on Urban Pest Management asked certain federal agencies to provide information on their urban pest control involvement or activities (see Appendix B). Although many instances of urban pest control involvement were reported, none of the agencies could point to any legislative authority for the development of urban pest control policy, or even to the development by the agency of a coherent and reasoned administrative approach to the subject.

U.S. Environmental Protection Agency

Federal Insecticide, Fungicide, and Rodenticide Act The U.S. Environmental Protection Agency (EPA) was established by executive order and Reorganization Plan No. 3 in 1970.[1] Its jurisdiction includes water- and air-pollution control, control of solid and hazardous wastes, control of noise, regulation of toxic substances, and under the Federal Insecticide, Fungicide, and Rodenticide Act (FIFRA), the regulation of pesticide use.

That act, (FIFRA),[2] as last amended in 1978, defines "pest" broadly, to mean:

(1) any insect, rodent, nematode, fungus, weed, or (2) any other form of terrestrial or aquatic plant or animal life or virus, bacteria, or other micro-organism (except viruses, bacteria, or other micro-organism on or in living man or other living animals) which the Administrator declares to be pests under section 25(c)(1).[3]

A pesticide is defined as a substance "intended for preventing, destroying, repelling or mitigating any pest," and further includes plant regulators, defoliants, or desiccants.[4]

FIFRA is concerned with "pesticide control" and, as a whole, does not deal with "pest management" (Grad 1971-1979). Indeed, that term, or any equivalent term, is not used in the act, except in Sections 4 and 20 where reference to "integrated pest management" is made. The law provides in considerable detail for the registration of pesticide substances, the suspension or cancellation of such registrations, and the limitation of pesticide use to registered pesticides. The law requires the EPA Administrator to register a pesticide if he determines that its composition warrants the proposed claim of efficacy, that its labeling and other information about it comply with the requirements of the act, that "it will perform its intended function without unreasonable adverse effects on the environment," and that, "when used in accordance with widespread and commonly recognized practice, it will not generally cause unreasonable adverse effects on the environment."[5]

The phrase "unreasonable adverse effects on the environment" is defined in the act to mean "any unreasonable risk to man or the environment, taking into account the economic, social, and environmental costs and benefits of the use of any pesticide."[6] The Agency's strong position in pesticide management is shown by the regulatory policy outlined in 40 C.F.R. Sec. 162.11 (1978) and termed the "rebuttable presumption against registration" (RPAR), which requires that any doubt as to whether conditions for registration have been met must be resolved against an applicant.

The act allows the Administrator of EPA to classify pesticides for general or restricted use, or for both general and restricted use where some of a particular pesticide's uses should be restricted. The restricted classification applies if the pesticide, when used in accordance with directions or in accordance with commonly recognized practice "may generally cause . . . unreasonable adverse effects on the environment, including injury to the applicator. . . ."[7] Such restricted-use pesticides may be applied only by "certified applicators" who are certified under a state or federal certification plan. To be certified, the "individual must be determined to be competent with respect to the use and handling of the pesticide, or to the use and handling of the pesticide or class of pesticides covered by such individual's certification"[8] (although, in what is perhaps a unique provision of law, competence is to be established by "his completing a certification form," with the state or the EPA Administrator expressly prohibited from requiring an examination to establish competency in the use of pesticides). It is worth emphasizing that the purpose of certification is not to establish competence in the management or control of pests but in the adequate handling of pesticides.

To the extent that FIFRA deals with the question of pest control at all, it is clear that the agricultural rather than the urban setting is emphasized. When cancelling the registration of a pesticide, or when changing its classification, the Administrator of EPA must take into account the impact of the proposed action "on production and prices of agricultural commodities, retail food prices, and otherwise on the agricultural economy." The Administrator must also notify the Secretary of Agriculture of any such proposed action. The requirement of notification may be waived only if the Administrator finds that suspension is necessary to prevent an imminent hazard to human health.[9]

The primarily agricultural orientation of the act is also demonstrated in Section 28, which requires the Administrator, "in coordination with the Secretary of Agriculture," to "identify those pests that must be brought under control." The Administrator is also to "cooperate with the Secretary of Agriculture's research and implementation programs to develop and improve the safe use and effectiveness of chemical, biological, and alternative methods to combat and control pests that reduce the quality and economic production and distribution of agricultural products to domestic and foreign consumers."[10]

FIFRA does not require coordination with agencies other than USDA, many of which have an interest or responsibility, or both, in urban pest management.

Resource Conservation and Recovery Act   The emphasis of the Resource Conservation and Recovery Act of 1976 (RCRA) is, as its name indicates, on conservation and resource recovery—including energy recovery—from wastes. RCRA is the direct successor of earlier federal laws that, beginning in 1965, exercised increasing federal control over the management (i.e., collection and disposal) of solid wastes. Earlier federal law was designed to provide technical and financial assistance to states and municipalities to allow them to perform more effectively what was viewed largely as a local responsibility to collect and dispose of residential, commercial, and institutional wastes. By 1976, however, when RCRA was enacted, the need for greater federal involvement had become apparent because of the integral connections between solid-waste management, air and water pollution control, protection of drinking water supplies, protection against environmental and health hazards from the disposal of toxic and hazardous wastes, and the growing shortage of urban land available for landfills.[11]

Although there is little evidence in federal solid-waste legislation— including RCRA—that pest management was a major consideration in the establishment of solid-waste programs, it is apparent that federal solid-waste legislation deals with a problem that involves major aspects of health and quality of life in cities. In many parts of the United States—even in urbanized parts—management of solid wastes did not become a problem until after World War II. Treated initially as a local problem of collection and disposal, the broader national implications such as resource conservation and land use did not emerge until well into the sixties. Then, as has been recounted in greater detail elsewhere,[12] federal interest in the field encouraged greater involvement by state governments, particularly in the regulation of solid-waste disposal. With full federal participation under RCRA, local, state, and federal agencies are now involved, with EPA assuming the dominant policy role, and state and local agencies are involved in both regulatory and service functions. The service function of collecting and disposing (or at least, of regulating the collection and disposal) of wastes has largely remained a local municipal function.

Although pest management was not a dominant concern when federal solid-waste legislation was enacted, better waste disposal programs clearly advanced the cause of urban pest management. The relationship of waste management to pest control, particularly with respect to rodents, cockroaches, and flies, is so well established as to need little comment.[13] In the urban setting "environment modification," an important aspect of integrated pest management, is almost synonymous with environmental sanitation and waste management and has the effect of depriving rats and insects of harborage and food. Thus, sound collection and disposal practices are essential pest control mechanisms, and environmental control over them is of the utmost importance in the urban environment.

RCRA provides a variety of approaches to the solid-waste problem. It establishes a federal Office of Solid Waste within EPA and creates Resource Recovery and Conservation Panels to provide technical assistance to states and localities. It also establishes a detailed federal program—with possible assumption of enforcement powers by the states—for the handling of hazardous wastes. It requires the development of state or regional solid-waste plans and conditions federal support on compliance with federal planning requirements and federal approval of state plans. As part of the process, the EPA Administrator issues federal guidelines for solid-waste plans, and the law establishes minimum requirements for the approval of plans. The requirements include phasing out all open dumps within five years from publication of an inventory of existing open dumps by the Administrator, and the replacement of dumps, at the very least, with acceptable sanitary landfills.[14] After consultation with the states, and after notice and public hearing, the Administrator must promulgate criteria for sanitary landfills to assure that there is no reasonable probability of adverse effects on health or the environment from the disposal of solid wastes.[15]

The suggested guidelines are to provide "a technical and economic description of the level of performance that can be attained by various available solid waste management practices (including operating practices) which provide for the protection of public health and the environment." The guidelines are also to describe the levels of performance necessary to protect subsurface and surface wastes so as to comply with applicable federal law, Clean Air Act requirements, and provide for "(E) disease and vector control; (F) safety; and (G) aesthetics. . . . "[16]

These guidelines and regulations reflect some awareness of the requirements of pest control, showing that EPA sought to carry out Congressional intent, as expressed in the House Report accompanying the RCRA in which reference is made to the adverse impacts of improper disposal of solid wastes, including "disease transfer (via such vectors as rats and flies)."[17] In proposing regulations outlining the criteria for sanitary landfills, EPA requires:

that the facility minimize the availability of food and harborage for disease vectors, and, when necessary, use other means to control disease vectors. Of particular concern are rodents. At facilities that dispose of uncomposted or unprocessed putrescible wastes, an effective means to control rodents may be the application of cover material at the end of each operating day. . . . [18]

There is a suggestion that there may be circumstances under which other means, including the use of rodenticides, are necessary. The proposed regulations provide considerable leeway for a variety of control methods, including integrated pest management.

U.S. Department of Housing and Urban Development

The U.S. Department of Housing and Urban Development (HUD) has a special interest in urban pest management because of the relationship of pests to places of human habitation. The multifaceted problem of blight in the inner city, with its symptoms of housing decay, filth, pests, and health problems, has long been recognized in federal legislation on slum clearance and urban renewal, and has been commented on in such eminent studies as the National Commission on Urban Problems ("Douglas Commission") Study, *Building the American City*.[19]

HUD's objectives with respect to pest management are delineated in various laws, particularly those that address slum clearance and urban renewal in inner-city areas. For example, grants are authorized from the Federal Government to "cities, other municipalities, counties and Indian Tribes . . . to assist in financing the cost of demolishing structures which under state or local law have been determined to be structurally unsound, a harborage or potential harborage of rats, or unfit for human habitation, and which such city, municipality, or county has authority to demolish." The law further requires that such demolition follow a plan to further the overall renewal objectives of the locality or "be consistent with a systematic rodent control program being undertaken in the neighborhood."[20] The law also requires, among other conditions, that the locality enforce local housing codes. Blighted areas can receive assistance for other actions that relate to pest management, including "the improvement of garbage and trash collection, street cleaning, and similar activities."[21]

Until 1975, HUD enforced the so-called "workable program" requirement of the Housing Act of 1949, which provided authority to make grants or loans for slum clearance and urban renewal on condition that the municipality had "a workable program for community improvement" and that it had "a minimum standards housing code, related but not limited to health, sanitation and occupancy requirements, which is deemed adequate by the Secretary, and (B) the Secretary is satisfied that the locality is carrying out an effective program of enforcement to achieve compliance with such housing code. . . . "[22]

The slum clearance and urban renewal title of the federal housing law was phased out in 1975, after Congress developed a new approach in the Housing and Community Development Act of 1974 (Public Law No. 93-383, 88 Stat. 633 [August 22, 1974]). Nevertheless, the 1949 act and its amendments have had lasting impact through the widespread adoption of housing codes in thousands of American cities; before the act, a mere handful had such codes. The "minimum standards" housing code referred to earlier was construed by HUD to refer to a number of national "model" housing codes, all of which contained provisions for the disposal of garbage and for control of rodents and insects, and all of which required close-fitting doors and other rat-proofing protection, and the screening of windows. Some of the codes also delineate the responsibilities of owners and tenants with respect to pest control inside and outside residences.[23]

The Housing and Community Development Act of 1974 continued the earlier approach of dealing with blight in general. Although no specific mention is made of pest management, the general authority provided to assist communities in dealing with urban blight is broad enough to encompass pest management as one method of general environmental improvement.[24] Funds are authorized for "code enforcement in deteriorated or deteriorating areas in which such enforcement, together with public improvements and services to be provided, may be expected to arrest the decline of the area."[25] Thus, HUD remains involved in activities that are inextricably a part of urban pest management.

In addition to its regulatory involvement in urban pest management, HUD plays an indirect role through Federal Housing Administration (FHA) mortgage insurance programs and multifamily subsidized housing,[26] and through the administration of HUD-owned and assisted housing.[27]

The HUD housing program encompasses about 4 million multifamily rental apartments and homes. Of these, 1.3 million are owned and operated by public housing authorities (PHAs), and over 500,000 are included in the Section 8 existing program for rental units. The rental units represent about one-seventh of all rental units in the country. The potential annual cost for pest control for all 4 million units could range from $40 to $80 million for management of the buildings involved, and an additional $20 to $40 million paid by tenants. The tenant cost is based on the knowledge that many families purchase their own pesticides to supplement management efforts and occasionally hire their own exterminators, even in public housing. There are no current figures on actual costs, nor does HUD indicate in most instances how pest control is managed.

In the FHA mortgage insurance program HUD is involved in protecting insured properties, whether urban or not, against wood-destroying pests. To protect the purchaser as well as its own interests, FHA requires soil treatment for protection against termites when legally permissible as well as the use of other methods to protect structures against decay and wood-destroying insects.[28] In multifamily rental housing programs that involve direct or indirect subsidy, HUD reviews management operations that may include pest control, though the Department does not become directly involved in daily operations. The responsible party, therefore, is state or local government, or quasi-public or nonprofit owners, or a private operator. In HUD-owned property, acquired through mortgage foreclosure and frequently referred to as "Secretary-owned," pest control work is done under performance contracts for pesticide spraying.[29]

HUD has historically had a large role in public housing programs. Its field staff has assisted PHAs in reviewing pest control operations and contracts. The Department is also updating its Public Housing Management Handbooks on Household Pest Control, Lawn Care, Rodent Control, and Termite Control, in view of the most recent requirements of the Federal Insecticide, Fungicide, and Rodenticide Act and to provide further instruction on integrated pest management. Where HUD has not provided direct assistance to PHAs, it can serve as a conduit. In Colorado, for example, HUD and EPA, assisted by the National Pest Control Association, developed an interagency agreement to implement a pest control training program for housing authority personnel.

Other agencies have also assisted HUD's clientele. A number of extension entomologists have assisted in training programs and in reviews of local PHA efforts. The U.S. Public Health Service and its local branches have assisted PHAs in reviewing or implementing rodent control programs (e.g., the National Capital Housing Authority, Washington, D.C., and the Atlanta Housing Authority) and USDA is assisting HUD in updating the Department's information on termite control.

HUD has included pest management as part of its training for field maintenance engineers, and entomologists from various universities and rodent control experts from the Public Health Service have assisted in the training. This training has emphasized integrated pest management and contracting with professional pest control operators (PCOs). Since it is generally accepted that a single-effort spraying program does not work well in public housing, more sophisticated programs are often called for.

222   URBAN PEST MANAGEMENT

Field engineers have been trained in the need for consulting entomologists to review PHA efforts and bid specifications for PCOs as well as in the use of professional service contracts for pest control firms.

There are no statistics on the extent of pest damage in HUD-insured or -subsidized housing, but in cooperation with the Bureau of the Census the Department does publish national data on rat and mice extermination services provided to owner-occupied and rental units. The most recent information available is for 1977 (U.S. Department of Housing and Urban Development 1977). These data do not include costs.

In summary, although the department's efforts in dealing with structural pests have generally been positive, there is no indication that HUD's efforts amount to a coherent programmatic policy on urban pest managment.[30]

U.S. Department of Health, Education, and Welfare

The interest of the Department of Health, Education, and Welfare (HEW) in urban pest management is primarily with respect to the prevention of disease. The Department has authority to make grants to the states for disease control programs and, as part of such grants, there are appropriations "for preventive health service programs for the control of rodents," thus continuing rodent control programs that would otherwise have expired under the general disease control authorization.[31]

HEW's interest in health is also reflected in the establishment of community health centers which include among their services (in addition to primary and supplemental health services):

environmental health services, including, as may be appropriate for particular centers (as determined by the centers), the detection and alleviation of unhealthful conditions associated with water supply, sewerage treatment, solid waste disposal, rodent and parasitic infestation, field sanitation, housing, and other environmental factors related to health, . . . [32]

The Food and Drug Administration (FDA) has a special responsibility in enforcing the Federal Food, Drug and Cosmetic Act, which has implications for urban pest management. FDA is responsible for the protection of much of the food consumed in the United States. Food is processed in or handled by some 77,000 commercial establishments, most of them located in cities or in urban areas (U.S. DHEW 1979). The Food, Drug and Cosmetic Act prohibits the movement in interstate commerce of adulterated or misbranded food, drugs, and cosmetics. Food is considered adulterated if it bears or contains poisonous or deleterious substances not required in its production or if it contains filthy, putrid, or decomposed substances or is otherwise unfit for human consumption. Moreover, food may be considered adulterated if it has been prepared, packed, or held under unsanitary conditions in which it may have become contaminated or rendered injurious to health.[33] Sanitary conditions require absence of pests or pesticides in food products except as permitted by FDA-established standards. FDA has been aggressive in its enforcement activities, and its actions in holding executives of food corporations personally accountable for violations of the law have had a major impact on pest management activities in food establishments.[34]

There still remains, however, a great deal of filth and pest infestation in the postharvest food supply in the United States. A survey by the U.S. General Accounting Office (1975), for example, indicated that, because of lack of money and personnel, FDA was unable to identify all the food plants operating under unsanitary conditions. FDA was requested to inspect 97 food manufacturing and processing plants with annual sales of about $443 million of bakery products, candy, cheese, ice cream, chips, spices, and other foods. Plants to be inspected were selected at random from about 4,550 plants in six FDA districts, including 21 states. GAO auditors accompanied FDA inspectors on 95 of the 97 plant inspections. The findings indicated that 39 (40 percent) of the plants were operating under unsanitary conditions, and of these 23 (about 24 percent) were operating under serious unsanitary conditions that had potential for causing or having already caused product contamination. From 1972 through the mid 1970s approximately 890 inspectors were hired by FDA as a result of GAO's findings. Budgetary limitations, however, have again forced FDA to reduce the number of inspectors and now a majority of inspections is carried out by the states. In some states the efficacy of these inspections is questionable. One FDA authority, in evaluating a continuing rodent problem in a food plant, commented that the building was still standing only because "the rodents probably all joined hands to help support it" (T.A. Granovsky, Texas A&M University, personal communication, 1979). If state or federal inspectors permit such levels of infestation to exist, the health and well-being of the general public and of the postharvest food delivery systems can be seriously affected.

FDA pest management policies as they relate to food establishments are not primarily urban pest management policies but are related to the special problems of the food industry. Because many food-handling establishments are located in urban areas, the FDA's actions in compelling better pest management have had an inevitable impact on the urban areas surrounding such establishments. FDA also has responsibility for setting tolerance levels for pesticide residues on new agricultural products,[35] another pest-control-related activity that is of general applicability.

U.S. Department of Agriculture

The U.S. Department of Agriculture (USDA) has no statutory authority to concern itself with pest management in urban areas. Its influence on urban pest management is indirect, in that it conducts a variety of programs that affect the management of pests in cities. The Forest Service, for instance, is responsible for research on wood-destroying organisms and has investigated the biology of termites, some beetles, and wood-destroying fungi.[36] This research emphasizes the use of persistent pesticides and is not integrated with USDA programs on other pests.[37] In addition, the Cooperative Forestry Assistance Act[38] authorizes the Forest Service to provide financial and technical assistance to the states for pest management in urban forests.

There are other USDA programs that have an impact on urban pest management. The Science and Education Administration's (SEA) agricultural research staff deals with insects affecting man and animals, stored products including grains and fabrics, and commodities being transported.[39] The SEA's cooperative research staff coordinates research on all such activities at state institutions. The USDA school lunch program[40] affects the transportation, storage, and serving of food, and the Federal Grain Inspection Service supervises the storage of grain and is responsible for protecting stored grain against the presence of pests and pesticides.[41] The Food Safety and Quality Service[42] is responsible for federal meat and poultry inspection programs.[43] Federal inspection certificates are not provided to facilities having unacceptable sanitation or inadequate pest control.

USDA acknowledges the need for special "research, action, and educational programs on pest management for urban audiences [which] will have to differ from those conducted for farmers[,]" (see working paper by Good, in Appendix B). In 1965 the USDA Extension Service made a grant of $38,000 for a study on "The Effect of a Planned Communication Program on Change of Attitude and Knowledge of the Urban Dweller Relative to Chemicals and Pesticides" (see working paper by Good, in Appendix B). USDA also sponsors a broad information and publication service, and some 42 publications are said to relate to the needs of urban residents for pest management information (see working paper by Good, in Appendix B). Most of these deal with the management of garden and ornamental tree pests or with pest management problems not directly related to urban environments. This is also true with respect to USDA public service television announcements and color slide sets, which are addressed to a general audience.

The crop protection research program of USDA's Science and Education Administration spent most of its funds on horticultural activities, but how much was addressed to urban horticulture is not known. It is acknowledged that, particularly with respect to integrated pest management, "There are staggering logistical problems in extending pest management information to urban audiences" (see working paper by Good, in Appendix B).

USDA states that for Fiscal Year 1979 SEA/AR (Science and Education Administration/Agricultural Research) planned to allocate $39,726,473 and 340.7 scientist-years of effort "in research relevant to, but not necessarily limited to, urban pest management" (see working paper by Good, in Appendix B).

USDA's Forest Service has no specific pest management program for urban forests, but during Fiscal Years 1978 and 1979 its Forestry Extension Program devoted some $5 million to Dutch elm disease control.

The Animal and Plant Health Inspection Service (APHIS) of USDA is charged with preventing the importation and spread of pests in the United States, a job that has become increasingly more difficult because of modern air transportation and the general mobility of people. Because air terminals are usually located near population centers, the activities of APHIS, though primarily directed at protecting domestic agricultural crops and animals, also have significant ramifications for the protection of public health.

USDA also cooperates with the U.S. Department of the Interior and with HEW in educational efforts relevant to urban areas, such as HEW's mosquito control projects and the Interior Department's Animal Damage Control and Wildlife Service information activities on rodents, snakes, bats, birds, and other vertebrate pests.

In addition, USDA cooperates with EPA and state cooperative extension services in conducting a pesticide applicator training program, but the Department acknowledges that this training program is primarily for farmers who apply "restricted use" pesticides and that urban home gardeners are not encouraged to take the course because the Department lacks the resources that might be required if the course was opened to the millions of urban gardeners.

USDA reports great success with the Master Gardener Program operated by many state cooperative extension services under the Smith-Lever Act of 1914. Begun in 1972 in the state of Washington, the program certifies "master gardeners" who then work as volunteers out of county extension offices advising home gardeners on numerous subjects, including pest control methods. And in a number of states the community development staff of USDA's Cooperative Extension Service has worked with local governments in establishing mosquito and rodent control programs.

In 1978 the state cooperative extension services devoted about 500 man-years to home horticultural education programs and about 600 man-years to the pesticide applicator training program, as well as substantial numbers of man-years to pesticide impact assessment and to IPM programs. How much of these efforts was devoted to urban problems is apparently not known, but it is known that state cooperative extension services are spending $3 million in federal funds annually to teach gardening to inner-city residents in 16 major cities. The goals of the program are to encourage low-income and minority groups to produce supplemental food, establish urban 4-H programs, and provide horticultural therapy for the elderly and the disabled. Pest management is part of the program but not one of its objectives.

USDA's SEA also administers a program conducted by state cooperative extension services to assist crop and livestock producers in developing IPM programs. The $5.4 million appropriation of Smith-Lever funds is also being used to develop urban IPM projects for lawns, gardens, ornamental trees, and homes, but the program is still in its early stages. Meanwhile, the USDA Extension Committee on Organization and Policy (ECOP) has published a study on *Integrated Pest Management Programs for the State Cooperative Extension Services* (USDA 1979), which addresses urban pest management under the heading of "homeowners and public health."

As the discussion above illustrates, many USDA activities are narrow in focus and not primarily directed at urban pest problems, and their effectiveness would seem to depend on integration with other urban pest management efforts. The working paper (see Appendix B) prepared for this Committee by the Department of Agriculture refers to a document called *Policy on Management of Pest Problems* issued by the Secretary of Agriculture in 1977, which places some emphasis on "the use of integrated pest management methods, systems, and strategies that are practical, effective, and energy-efficient." That policy document also includes the following statement: "[I]n carrying out its pest management policy, the Department will be mindful of the interest and needs of all segments of society, including those interested in households, gardens, small farms, commercial farms, forestry, food and fiber handling, transportation, storage, and marketing enterprises." Although the policy document is broad in scope, it does not enunciate an urban pest management policy.

The working paper prepared by USDA cites a 1978 Gallup poll which showed that 41 percent of U.S. households had a vegetable garden and emphasized the need for better IPM programs in urban and suburban areas as well as in small and medium-sized communities. The working paper calls attention to urban needs as follows:

In the urban setting there are two large groups that need assistance. Research, action, and education needs are unique for these groups, each requiring different resources and organizational relationships. The largest group includes individual homeowners with problems relating to lawns, ornamentals, shade trees, gardens, and household pests. The other group includes urban institutions such as city and county park commissions, schools, hospitals, golf courses, and cemeteries.

The Committee's interpretation of this statement is that USDA's view of the nature of urban pest problems is severely limited, even though the working paper indicates that the Department—while meeting its primary obligations to farmers—is also doing work that benefits urban pest management. But the working paper indirectly documents the point made earlier, namely, that no federal department or agency is responsible for urban pest management. Although USDA's contribution should not be deprecated, the Department's direct involvement in urban pest management seems minimal given the full dimensions of the problem.

Other Federal Agencies

Other federal agencies also play an indirect role in urban pest management. The National Park Service of the Department of the Interior, for instance, is responsible for maintaining several urban parks and buildings and for the preservation of many historical sites.[44] Interior's Bureau of Indian Affairs develops specifications and contracts for pest control services at schools, warehouses, office buildings, and other structures used by American Indians[45] in areas that may be regarded as urban.

Interior's Fish and Wildlife Service performs research and makes recommendations on vertebrate pests and their control. Its Animal Damage Control Division[46] has expertise in the management of rats, mice, bats, birds, and a variety of other vertebrate pests, some of which cause problems in urban settings.

The General Services Administration (GSA), which is responsible for the design, building or leasing, operation, and maintenance of federal buildings, is involved in urban pest management in two ways.[47] GSA has its own pesticide applicators and also contracts with private pesticide applicators to provide pest control services in its buildings. In addition, GSA influences the selection of pesticides used in federally operated buildings through its Federal Supply Service.[48]

The Veterans Administration (VA), which guarantees repayment of mortgages extended to veterans, requires inspection of homes purchased by veterans to determine structural soundness and the possible presence of wood-destroying pests.[49] In addition, the VA operates a pest control service for its hospitals and other buildings.

Significant work related to urban pest management is also done by the Department of Defense (DoD). DoD supplied the Committee with an extensive working paper (see Secretariat, Armed Forces Pest Management Board, in Appendix B) on its operations that indicates that the Department's pest management program costs approximately $110 million a year. The program is designed to protect 3 million civilian and military personnel as well as equipment and a variety of subsistence items and wooden structures valued at $100 billion, located on 26 million acres of real estate worldwide. DoD engages in substantial research and has an elaborate organizational structure for pest management. The Armed Forces Pest Management Board is responsible to the Deputy Assistant Secretary of Defense and to the Assistant Secretary of Defense for recommending policy, serving as a scientific advisory board, coordinating technical and scientific pest management activities, operating a pest management information analysis center, and identifying pest management research requirements. DoD also has an EPA-approved plan for certification of pesticide applicators and has established a common definition of integrated pest management throughout the Department. The DoD certification plan is the only EPA-approved federal certification plan, and DoD maintains a pest management force of approximately 3,200 full-time civilian and military personnel.

It is clear that DoD's pest management problems are worldwide and substantial, and that with 3 million military and civilian employees, DoD facilities are subject to many of the same problems found in high-density urban and suburban environments. Moreover, military installations are often situated close to civilian communities. As a result, DoD pest management activities have significant implications for these communities.

DoD classifies urban pest problems as either industrial institution-based or related to personal and public health. The Department uses IPM techniques and clearly has a great deal of experience in pest management in high-density areas as well as in such unusual sites as ships at sea. Although DoD's experience in all these areas may be somewhat applicable to urban pest management, it is apparent that DoD has a somewhat different problem in managing pests from that faced by civilian governments. The nature of DoD controls and the military chain of command probably assure more immediate and direct response to pest problems than can be expected in most urban areas. DoD buildings and installations in urban areas may present pest problems that require coordination with municipal efforts, and under some circumstances the Department may assist nonmilitary communities in solving pest problems, but beyond this DoD has no responsibility to help solve nonmilitary pest problems.

The Council on Environmental Quality and the President

The Council on Environmental Quality (CEQ) is preparing a major study of integrated pest management. The study had not been released at the time of the completion of this report. CEQ was established by the National Environmental Policy Act of 1969[50] and is an advisory body responsible to the President. Although it has no administrative or operational obligations—other than to promulgate regulations (earlier, "Guidelines") for the preparation of environmental impact statements under the National Environmental Policy Act[51]—its advisory role is significant, and its report on IPM may well have significant implications for urban pest management.

President Carter indicated support for integrated pest management in his 1979 environmental policy message, which, after noting the benefits of integrated pest management in the light of pest resistance to pesticides, provided the following policy directives:

The Federal government—which spends more than $200 million a year on pest control research and implementation programs—should encourage the development and use of integrated pest management in agriculture, forestry, public health, and urban pest control. As a result of a government-wide review initiated by my 1977 Environmental Message, I am now directing the appropriate federal agencies to modify as soon as possible their existing pest management research, control, education, and assistance programs and to support and adopt IPM strategies wherever practicable. I am also directing federal agencies to report on actions taken or underway to implement IPM programs, and to coordinate their efforts through an interagency group.[52]

Subsequently, the President established an interagency IPM coordinating committee consisting of the Secretaries of Agriculture, Commerce, Defense, HEW, HUD, Interior, Labor, Transportation, the Administrators of EPA and GSA, and the Chairman of the Council on Environmental Quality to assure implementation of the directives contained in his message to Congress and to oversee further development and implementation of integrated pest management practices.[53]

THE STATE RESPONSIBILITY

Many state programs of environmental control and management have often followed federal requirements and models ever since the Federal

Government—and particularly EPA—became the principal protector of the environment. In many instances the Federal Government initiated the protective legislation, while in others the federal legislation required implementation through a state plan or required compliance with federal law or criteria as a condition for obtaining federal grants-in-aid.[54] This is the case in both the regulation of pesticides and the regulation of pesticide applicators,[55] as well as in solid waste management, which has important implications for urban pest management.[56]

There is a substantial area in which federal regulation has preempted state control of pesticides. The Federal Insecticide, Fungicide, and Rodenticide Act (FIFRA) is fully preemptive with respect to labeling or packaging of pesticides; the states are barred from imposing or continuing in effect any requirements relating to that subject "in addition to or different from" the requirements of FIFRA.[57] States may, however, regulate "the sale or use of any pesticide" if state regulations do not permit a sale or use prohibited by FIFRA.[58] Thus, the states may impose requirements that are at least as stringent as federal requirements. The states are also free to provide registration for additional uses of federally registered pesticides that are formulated for distribution and use within the state to meet special local needs. Under this provision, a state registration is effective unless the Administrator of EPA disapproves it within 90 days after its issuance. The special state registration allows distribution of the pesticide only in the state of registration, and if disapproved by the Administrator within the 90 days, it will not be effective for longer than that period.[59]

Although FIFRA contains no specific provisions on local regulation of pesticides, the legislative history of the act clearly indicates that Congress intended to bar local pesticide regulation.[60] Thus, it would be invalid and impermissible for a state to pass legislation purporting to grant such authority to municipalities or other local governments.

As mentioned earlier, FIFRA requires the Administrator of EPA to set standards for the certification of pesticide applicators.[61] To be certified, an applicator must be competent in the use and handling of pesticides generally, or in the use and handling of specific pesticides or class of pesticides covered in the individual's certification. After the Administrator has set the standards, the states, if they desire to certify applicators, may submit a state certification plan to the Administrator. In submitting a state plan, the Governor must designate a state agency that will be responsible for its administration and must also demonstrate that the agency has the legal authority, qualified personnel, and adequate funds to carry out the plan.[62]

Many states have taken advantage of the opportunity to certify applicators. The 1975 amendment of FIFRA made it significantly easier for applicators to become certified by providing for unsupervised self-certification. The law now provides that under a state plan the certification standard shall be "deemed fulfilled" by an applicator "by his completing a certification form." Although the Administrator "shall further assure that such form contains adequate information and affirmations to carry out the intent of this Act, and may include in the form an affirmation that the private applicator has completed a training program approved by the Administrator," the requirement is rendered less than compelling by the further language that completion of a training program may be required "so long as the program does not require the private applicator to take . . . any examination to establish competency in the use of the pesticide."[63]

The law also authorizes the EPA Administrator to require pesticide dealers participating in a certification program to be licensed under an approved state licensing program.

After the Administrator has approved a state plan for the certification of applicators, periodic reports may be required from the state agency. If the Administrator does not approve a plan he must give notice to the state, which can request a hearing.[64] Similarly, if the Administrator determines that an approved plan is not being implemented properly, the state is given notice of this disapproval and may request a hearing. Approval can be withdrawn if the state does not take "appropriate corrective action" within 90 days.[65]

In general, state legislation regulating pesticides parallels federal law, though not all states have as yet caught up with the requirements of FIFRA. Thus, some states still follow FIFRA as amended in 1954 and do not regulate use of pesticides in keeping with the present federal law. Some states, in spite of federal preemption of the field, still have provisions in their laws that provide for labeling of pesticides, including requirements that the registrant assume responsibility for the consequences of the "commonly recognized use" of the pesticide. Labels usually contain a list of ingredients and the pesticide registration number. In view of the fact that FIFRA is preemptive with respect to federal labeling and packaging requirements, some state laws may have been partially invalidated by FIFRA.[66]

Following the amendment of FIFRA in 1972 it became necessary for the states to revise their pesticide control laws to make them conform to the requirements of the new federal law. Accordingly, the Council of State Governments joined with The Association of American Pesticide Control Officials (whose members are responsible for enforcing state pesticide control laws) and with members of industry to cooperate in the preparation of a new model act. EPA also was involved because of the need to coordinate federal and state programs, as provided for in the federal law. A model Pesticide Use and Application Act was prepared and appeared in the 1974 *Suggested State Legislation* volume of the Council of State Governments. The model has been used for many recent revisions of state pesticide laws to bring them into conformity with federal requirements.[67]

In cases where the state exercises control over the use of pesticides, state law closely parallels FIFRA. Generally, the head of the state agency that has the authority to control pesticide use will determine the classification of "restricted use" pesticides, after public notice and public hearing.[68] When a pesticide is so classified, its sale, purchase, or use is restricted to licensed persons. The department regulating pesticide use usually has authority to establish procedures for the certification of pesticide dealers and pesticide applicators.[69] Some states have a pesticide advisory board whose members are drawn from state environmental protection agencies and from the agriculture departments of state universities. The purpose of these boards is either to advise the department that regulates pesticide use[70] or to take a more active role in the regulatory process.[71]

Coordination of federal and state law has become even more important since the 1978 amendments of FIFRA, which grant the states primary enforcement responsibility for pesticide use violations if the state has adopted adequate pesticide use law and regulations and is implementing adequate procedures for their enforcement.[72] This delegation of enforcement powers is in addition to state responsibility for enforcing restrictions on pesticide use by certified pesticide applicators pursuant to cooperative agreements with the Administrator of EPA.[73] The Administrator retains full enforcement responsibility, however, in states that do not assume such primary enforcement responsibility and may resume the exercise of enforcement responsibility if it is found that the state is not carrying out its obligations properly[74] or that emergency conditions require immediate action because a state is unwilling or unable to respond adequately.[75] There appears to be no way of assessing at this time the effectiveness of the delegation of responsibilities, but it is clear that the entire machinery—just like the federal structure—was primarily designed for regulating agricultural pesticide use.

Many states have joined an interstate compact for the creation of an insurance fund to provide funds for eradication and control measures in a state threatened by a pest from outside its borders and to prevent the spread of pests from one state to another.[76]

There is relatively little in state law that deals explicitly with urban pest management. Many states authorize the establishment of mosquito control agencies or "mosquito abatement districts," and some delegate mosquito control authority to municipal and other local governments.[77] Control of rodents and insects is commonly delegated by state law to local health agencies or to local agencies in charge of enforcing housing standards. The delegation may be express, or it may simply be implicit in the general authority to protect public health and to prevent the spread of communicable disease, frequently contained in laws that generally authorize the abatement of public health nuisances.[78] Actual standards and programs are sometimes contained in state sanitary codes[79] but more often are established by local ordinances or local housing and sanitary codes.[80] In some states "vector control" is divided between state and local health agencies.[81] The common reliance on local regulation is implicit, too, in the "workable program" requirement under federal urban renewal and slum clearance legislation and in the federally aided code enforcement (FACE) program.[82]

Federal legislation on the collection and disposal of solid waste has stimulated a great deal of state legislation. There was virtually no state solid-waste legislation before the first federal law on the subject was passed in 1965. Since then, state legislatures have reacted to each new federal act, most recently to the 1976 Resource Conservation and Recovery Act, and every state presently has solid-waste legislation. The legislation covers a variety of matters, but, in general, state laws on solid waste deal primarily with methods of disposal, such as the use of landfills or incineration. The state response to federal legislation was largely based on the requirement that each state designate a single state agency to be responsible for the execution of state law and to submit a solid-waste management plan.[83] The state agencies responsible for solid-waste management are either public health or environmental control agencies, although in some states the task is divided between the two.[84] Normally, state legislation relating to landfills includes requirements relating to pest control, particularly rodent control. The development of state laws on disposal has had the natural consequence of involving the states in other aspects of solid-waste management, including the collection and transportation of wastes, that had previously been the sole responsibility of local governments. Most frequently, state legislation in these areas authorizes or requires action by municipalities and other local governments. Local governments are usually free to undertake waste collection as a government function or to follow a variety of other methods, including licensing of waste disposal operations, franchising of such operations, or contracting for municipal waste disposal with private firms.

Waste collection and disposal activities are still primarily a local government task.[85] In performing the task under appropriate state waste disposal plans, however, local governments must generally follow federal guidelines if federal support is sought by the state or locality, which is commonly the case.[86]

THE LOCAL RESPONSIBILITY

The Committee's assessment of local responsibility in urban pest management is based on (1) a review of selected local codes and ordinances and state laws pertaining to pest management issues and decision making (see Appendix E for list of states and cities that provided information), and (2) an examination of the problems and needs of pest managers in selected urban locales. This section is also supported, in part, by two working papers prepared for the Committee. One paper (see Weiner and Sapolsky, in Appendix B) presents case studies of local pest management programs in Boston, Massachusetts; Providence, Rhode Island; and Baltimore, Maryland; and the other examines inner-city participation in environmental management and planning (see Henderson, in Appendix B).

Legal Aspects of Local Pest Management, with Special Reference to Large Cities

In most of the legislation analyzed by the Committee, the local health department (either city or county, or both) was the agency primarily responsible for pest control. (Even when the health department was not primarily responsible for the enforcement of pest control ordinances, it was generally responsible for dealing with pest infestations that threatened public health.)[87] In only 2 of 40 local governments was the housing department involved in pest management.[88] In the case of entirely public activities—public buildings, schools, hospitals, transportation systems, urban renewal programs, public housing, etc.—pest management was ordinarily carried out as a part of the management of those facilities and programs, and was frequently outside the control of local regulatory agencies.

Other public administration units with pest management responsibilities included buildings departments, environmental protection agencies, and departments of public works. In larger cities, building code enforcement units have important pest control programs associated with the management of multifamily housing, and they sometimes addressed problems associated with abandonment, foreclosed property, and the like. Departments of public works often have pest management responsibilities for exterior public sites and transportation facilities and may undertake other projects, such as weed control. Environmental protection agencies usually represent a consolidation of previously existing activities, corresponding most often with integration of sanitation, sewer, and water supply activities, and sometimes with others such as air quality or resource recovery programs.

There was relatively little reference to state or federal law in the ordinances analyzed.[89] Several indicated that the state was responsible for pesticide regulation and enforcement. Several also indicated that the local health department operated under the authority of state enabling legislation.[90]

Private Responsibility for Pest Control   Under local pest control ordinances,[91] as a general rule, the "person in control" of private property is responsible for pest control. The New York City health code, for example, defines "person in control" as "the owner or part owner of a building, lot, premises, or commercial vehicle, the agent or occupant of a building, lot or premises or any other person who has the use or custody of the same or any part thereof."[92] Other codes are drafted even more broadly, wording the responsibility in terms of "any person."[93]

Thus, in the case of a single-unit dwelling, the owner (or the lessee, if the lease specifies that the lessee is responsible for maintenance and repairs) is responsible for the control of pests on the premises.[94] The owner typically is responsible for rat- or insect-proofing[95] in multiple-unit dwellings and commercial buildings, and for pest control in communal, shared, or public areas.[96] The owner is also generally responsible for maintenance of the rat-proofed or insect-proofed condition.[97] In the case of commercial buildings, however, the owner may be responsible for initial rat-proofing while the manager or occupant in charge of the building is responsible for maintaining that condition.[98] In general, the duties imposed on owners of business buildings are less stringent than those imposed on the owners of multiple-unit dwellings.

The occupant of a unit in a multiple-unit dwelling or commercial building is responsible for pest control on the premises.[99] But if a multiple-unit dwelling is generally infested, extermination is usually the responsibility of the owner, not the individual occupant.[100] This follows from the owner's duty to rat-proof or insect-proof, since a general infestation indicates failure to do so or to maintain that condition.

The common method of enforcement, as local ordinances indicated, is for the director of health to issue a notice to abate the nuisance. If the order is not complied with, one of two types of sanctions is generally employed: Either the local health department abates the nuisance itself and recovers the cost plus any penalties from the responsible party (either by court action or by putting a lien on the property),[101] or the ordinance provides that failure to abate the nuisance is a misdemeanor and that fines or imprisonment may be imposed against the responsible party.[102] In addition, if the infestation is so great as to present a serious threat to health, the director of health may be empowered to order the building closed until the infestation is eliminated.[103]

Target Organisms Very few of the local ordinances examined contained a comprehensive definition of the term "pest."[104] Most (29 of the 40 examined) considered rats a pest. Next in frequency were mosquitoes (22); flies (15); lice, ticks, mites, bedbugs, or fleas (12); cockroaches (11); pigeons (10); mice (8); and bees, wasps, hornets, or ants (6). A few mentioned spiders or animals that are potential carriers of rabies (dogs, bats, squirrels, skunks, and raccoons).

Although most localities had specific rat control ordinances and some had mosquito control ordinances, some localities dealt with all pests other than rats under general nuisance ordinances.[105] The health regulations of Fulton County (Atlanta), for example, define nuisance as "whatever is dangerous or detrimental to human life or health and whatever renders or tends to render soil, air, water or food impure or unwholesome."[106]

Other Considerations Some local ordinances specified in great detail the materials and construction techniques to be used in rat-proofing,[107] or the types of refuse containers allowed,[108] while others defined rat-proofing in general terms but left promulgation of specific standards to the Director of Health.[109]

Regulation of the collection of wastes, which is authorized by state law, clearly has implications for the way in which householders and commercial establishments store wastes prior to collection. With respect to household wastes, local ordinances on the collection of wastes and housing codes that call for sanitary storage of waste and specified waste containers both provide significant environmental controls. It is worth emphasizing here that the ultimate performance of many of the tasks that ensure proper waste management is carried out at the local level in spite of the increasingly greater involvement of state governments in this area.

Many local laws outlined a significant educational role for the local health department. Some examples are Fulton County's block sanitation program, which consists of door-to-door inspection and enforcement efforts within a specified area; Detroit's educational-enforcement pilot project in a 700-block area; and Indianapolis-Marion County's use of health aides to work on a one-to-one basis with those who contribute to or are affected by rodent problems. Other examples included mass media, school presentations, community meetings, distribution of pest control literature, and technical assistance to groups and individuals.

Administration and Management Issues in Urban Pest Control

Pest management is a significant and growing part of environmental activities carried out by local governments. Major cities allocate several hundred thousand dollars annually and employ several hundred people in pest control programs. The Committee examined local pest management practices for several reasons: (1) we felt strongly that effective change can only be carried out by local governments—i.e., where decisions are made to budget for and implement the details of such innovative approaches as IPM; (2) we were concerned that a federal policy, if implemented, be coordinated with local management capability; and (3) we recognized the need to understand the constraints and opportunities in the local management system. The Committee's examination of selected programs revealed that local pest management is carried out in a context of differential and uneven standards, fragmented activities and responsibilities, limited resources, and traditional philosophies and attitudes about how pest control ought to be handled.

Urban pest control may be viewed as a problem that is handicapped by old solutions and approaches. The problems of developing programs and activities to meet contemporary standards reflect the classic dilemma of public needs, demands, and expectations pitted against diffused governmental responsibility. Thus, when assessing environmental management problems where everyone expects to benefit but no one is conscious of cost, the public system automatically becomes fragmented (Ingram and McCain 1977). Public pest management further suffers from lack of support and innovation at a time when public expectations and standards are high. In addition, the perceived low status of pest control programs reinforces diffused responsibility and thus public bureaucracies are unlikely to compete for and promote activities which lack prestige and influence and where growth can be seen as an indictment of program effectiveness. This is an ironic distinction that pest management shares with other government activities such as public assistance, where budget growth is considered program failure or ineffectiveness rather than success.

The position or status of a program or activity within a local government is often a useful indicator of the priority given to specific problems and of the amount of resources devoted to their solution. Allocation of resources reflects a combination of pressure from constituents, perceptions of political leaders about public desires, internal bureaucratic values, and the professional biases of public employees. In an effort to understand how these factors influence urban pest management at the local level, the Committee examined the pest management programs of a number of municipalities and urban counties (see Appendix F) and made a site visit to New York City. Specific attention was given to such factors as manpower, materials, capital, equipment, and supervision. The findings summarized below are based primarily on responses from 30 of the 60 managers whose programs were examined in detail.

The Management Setting: Program Organization and Problems   Reform in local pest management programs tends to emphasize the comprehensive grouping of related activities into a superagency format. Two cases were found: (1) pest management integrated into an environmental management organization; and (2) pest management integrated into a housing and community development organization. In the first case, interaction is facilitated between pest control and related environmental maintenance activities such as litter control, solid waste management and disposal, sewer cleaning, and the like. This might assist in coping with problems such as tradeoffs between air quality and pest propagation with, for example, closing of municipal incinerators or dumps. A somewhat different goal might be achieved in the second case; internally, gains might be improved coordination with building code enforcement programs, while gains external to the organization might come in the form of more neighborhood-oriented pest management strategies.

Apart from the establishment of environmental protection agencies and housing and community development agencies, the Committee found little evidence of attempts to reform the structure of local government to deal with pest control. (One notable exception, however, was the creation in 1979 of a Department of Rat Control in Chicago.) Whether reform in agency programs or authority accompanied these changes is not known, but the consolidation of separate agencies into an environmental protection department may reflect attempts to introduce innovative planning and new technologies into the administration of traditional activities, such as solid waste disposal. Housing and community development activities have become strongly oriented toward "neighborhood" conservation, with housing and other environmental and public service issues taken more fully into account. Although a full evaluation of the organizational reforms found during the Committee's investigation is beyond the scope of this study, they may represent significant innovations in handling pest problems at the local level.

The effectiveness of local pest management as well as other public service activities is strongly influenced by government organization, the structure and quality of basic inputs, the distribution of responsibility, and coordination. In examining the structure of selected pest management programs, the Committee observed several features that may handicap the present system: (1) the existence of two separate management systems, one for private property and activities and the other for public management activities; (2) lack of innovative standards and technology transfer capabilities; and (3) a generally high level of expenditure with little or no effectiveness assessment.

The existence of two management systems appears to reflect the domination of local pest control activities by a rigid tradition and philosophy emphasizing environmental health over broader quality of life and non-health impact issues. Following this tradition, local pest management may be characterized as:

(1) enforcement and "evidence" oriented, particularly emphasizing site inspections for evidence of abuse rather than conditions contributing to problems; and

(2) discretionary prioritizing or setting of target site categories for major enforcement efforts.

The obsession with "evidence" forces local inspection programs to concentrate on such narrow evidence as "rat droppings," "presence of cockroaches," or "structural damage by vermin." While these simple criteria may help to reduce complex problems to manageable dimensions for coding and processing of information, they communicate little about the relative seriousness of the problem, the likelihood that the problem will persist, or other factors such as poor maintenance conditions that contribute to violations.

Prioritizing specific site categories may be done by statute, but more frequently, however, the decision reflects administrative choice and an effort to reduce program scope to manageable dimensions. Prioritizing site categories for enforcement purposes is widespread. Of 23 possible site categories, local emphasis was overwhelmingly in the food-retailing, food-processing, and residential common spaces such as hallways, cellars, and yards. By contrast, far less attention was given to health-care facilities, hotels, dormitories, and non-residential sites and activities. In narrowly targeting specific activities, the broader issue of quality of life tends to get ignored and inspection programs have been criticized on the grounds that they are generally ineffective. Beginning in the 1960s, it was found that, with few exceptions, the programs addressed symptoms rather than causes, used inadequate enforcement measures, and thus failed to change poor management and maintenance practices in both the public and private sectors in urban and slum environments (Bergman 1974; Field and

Rivkin 1975; Seidel 1979; Ventre 1971; and U.S. Congress, House 1968). Yet despite condemnation, traditional inspection programs continue to dominate local programs.

The role of the public sector in pest management is large because of the amount of public property in typical urban areas. At least one-third of the urban land mass may be in the public domain. In some cases, when undeveloped land, harbors and rivers, rail sidings, terminals, and yards are taken into account, the public land mass may exceed 50 percent. Local pest management activity concerned exclusively with public sites focuses on areas such as infrastructure and buildings, rights-of-way, rail property, and vacant land (in and adjacent to waterways in particular). The critical feature of this system, however, is that pest management: (1) occurs as a secondary or by-product of other activities such as highway maintenance, health care, education, public housing, etc.; and (2) is a function of a particular agency that sets standards, trains manpower, organizes resource input, and the like.

The Committee found little evidence of uniformity in standards for pest management practices, targets for management, or innovation. The typical process can be defined as broadly discretionary, but consisting of the following activities:

(1) identification of problems—either by maintenance crews, service users, or observers;
(2) development of consensus following numerous reports of problems; and
(3) decisions following several types of evaluations emphasizing potential inputs and costs.

The exact course of action adopted will vary considerably and depends on the availability of resources, the seriousness of the problem, and the existence of other priorities. Future budgets may be justified on the basis of need for additional pest management activities. Alternatively, agencies may elect to absorb additional costs in existing programs and budgets.

Innovation and reform are extremely rare in local pest management programs. Of the 60 programs examined, nearly 40 reported no recent change in legislative or other reform, and the same number reported traditional structures with primary responsibility divided between public health agencies and environmental management agencies. Despite the rapid shift in governmental reorganization in environmental activities (Haskell and Price 1973), few pest control activities appear to have benefited.

Data on complaints about pests served to identify the workload facing local pest managers. Local managers indicated that urban environments provide a short list of problem species. Rats, mice, cockroaches, fleas, flies, and mosquitoes led the list for most communities with only slight variations between regions and little evidence of variation because of community size, age, wealth, or socioeconomic characteristics. Sites generating problems presented a somewhat different picture. The three site categories mentioned most frequently included (1) multi-family housing; (2) open land; and (3) food-handling facilities. Other sites mentioned included single-family housing; hotels; prisons; port, harbor, and other transportation facilities; restaurants; retail food stores; hospitals; industrial sites; public buildings; and schools.

Budgets, Expenditures, and Staff A key ingredient in administering local pest management programs is the resource base and budget support devoted to public pest control activities. There are few federally sponsored categorical grants or grants-in-aid specifically devoted to local pest management activities. This leaves localities dependent on local tax resources or on discretionary (noncategorical) federal funds.

A considerable amount of local resources is devoted to pest management, with most large cities spending in excess of $500,000 annually (see Table 5.1). In examining a sample of 30 localities for which budget data were reported, the Committee found that the highest per capita expenditure ranged from $.50 to $.99 (see Table 5.2). This compares with expenditures reported for other environmental areas, such as solid-waste disposal, waste water treatment, and air quality.

Local pest control expenditures appear to be considerable because, as indicated in Table 5.1, many cities contribute considerably more than the 50 percent matching funds required for receiving federal assistance. Most of the federal grants are limited, however, to rat control. Localities also use a variety of other sources of funds, such as flexible grants under community development programs, revenue sharing, and special programs such as the Comprehensive Employment and Training Act (CETA) to augment existing resources.

One might assume that the highest per-capita expenditures for pest control occur in larger municipalities and that counties and smaller municipalities spend less, even when federal funds are not counted. This, however, was not a finding of the Committee's survey. There was no clear correlation between population size and the ratio of state-local to federal expenditure among the selected localities for which data were available (see Table 5.1). Caution must be used in interpreting the data, however. Some local expenditures may include state pest management funds, while federal support may include federal grants other than pest control grants, e.g., general revenue-sharing funds, community development block grants, and CETA and U.S. Public Health Service grants.

URBAN PEST MANAGEMENT

TABLE 5.1   Local Pest Management "Resource Effort" for Selected
United States Municipalities

| Municipality | Pest Control Expenditures (1979 dollars) | | | State-Local Effort (% of total) |
	Total	State-Local	Federal	
Akron	$375,000	$258,000	$ 67,000	68.80
Baltimore	105,630	630	105,000	NA
Chicago	6,079,235	2,578,725	3,500,510	42.42
Cincinnati	417,743	252,743	165,000	60.50
Cleveland	531,254	265,517	265,717	49.98
Des Moines	100,000	60,000	40,000	60.00
Detroit	2,800,000	2,450,000	360,000	87.50
Kansas City, Mo.	386,958	284,444	102,514	73.51
Louisville	530,032	365,232	164,800	68.91
Milwaukee	503,665	138,665	365,000	27.53
New Orleans	262,000	152,000	110,000	58.01
New York City	7,000,000	3,650,000	3,350,000	52.14
Philadelphia	4,660,000	3,466,000	1,200,000	74.38
San Francisco	250,000	107,000	143,000	42.80
St. Louis	1,002,276	509,805	492,471	50.86

NA indicates not available.

SOURCE: Based on NRC Committee on Urban Pest Management study of local pest
management programs.

TABLE 5.2   Amount Per Capita Spent on Pest Control by 30 United States
Municipalities (1979 dollars)

| Expenditure Per Capita | City Population | | | | |
	Million or More	Between 600,000 and 1 Million	Between 300,000 and 600,000	Between 100,000 and 300,000	Total
$2.00 or more	2	1	—	—	3
1.00-1.99	2	1	3	1	7
0.50-0.99	2	4	3	1	10
0.25-0.49	1	1	3	—	5
0-0.24	1	—	2	2	5
Total:	8	7	11	4	30

SOURCE: Based on NRC Committee on Urban Pest Management study of local pest
management programs.

Local staffing patterns also showed considerable variation. In several localities, such as Cincinnati, Cleveland, and Milwaukee, seasonal and part-time staff were used to supplement permanent staff. CETA workers are also used in several cities to assist regular pest management staff.

Deficiencies, Needs, and Innovations Perceptions of local managers about their programs suggested a less than adequate overall effort in local pest management. Much of their criticism focused on issues unrelated to expenditures, however. Although many (nearly 34 percent) of the managers felt that expenditures for local pest management should be increased, the majority ranked public education, stronger enforcement power, better training for personnel, and coordination with other agencies as most critical. Although community environmental standards have risen sharply in recent years, the ability of local governments to translate public demand into management and control rather than legal rules has been slow. There is a particular need, as one manager recognized, for "actual on-site efforts at management change and less empty laws." An assistant chief of the Vector Biology Control Section of the State of California Health Services stated: "There is a lack of programs directed toward prevention of conditions that are responsible for urban pest management problems" (E.W. Mortonson, California Department of Health Services, personal communication, 1979). A similar view was echoed by the Program Coordinator for Vector Control in New Jersey who suggested that major problems were not budgets, but more trained personnel and education (D. Adam, Vector Control, Trenton, New Jersey, personal communication, 1979).

There is additional evidence, however, that the need for innovation in local pest management programs extends beyond the issue of growth. For example, a recent study suggests that local management programs may not be sufficiently broad to capture a variety of consumer issues which fall outside the "private property inspectional system" or the public management system, but where impact is critical (Olkowski and Olkowski 1978).

Effective delivery systems and, in particular, treating the source of the problem are required. The program-management requirements for these broader approaches however, are far, far different from existing narrow concepts of need. Research in urban IPM technology transfer, from the IPM specialist to the political and maintenance personnel of the system to be managed, must take on a distinctively interdisciplinary approach. The ecologist-IPM specialist finds that incorporation of techniques of analyses and integration, from such varied disciplines as sociology, psychology, political science, public education and business management, becomes a necessary requirement. . . . The programs developed . . . so far have all had three major components: delivery system, education and research. . . . (Olkow-ski and Olkowski 1978).

Thus, in organizing a pest management system, technology transfer would require a totally different set of skills and training, different management philosophy, and a new flexibility to respond to and adjust to particular site or need requirements.

Restructuring the existing management framework to obtain higher levels of managerial control appears to be essential if the high degree of diffusion and differential standards and resources are to be controlled. Above all, there is a need for central management authority to set standards for both the private and the public sector and to have both the policy authority and the management tools necessary to carry out new and innovative approaches. Currently there is no means to assess the effectiveness of a particular program. Attacking the problem instead of the symptom, however, automatically creates standards for measurement; and reduction of environmental conditions conducive to pest propagation will automatically result in fewer pests.

PROBLEMS OF INFORMATION TRANSFER AND COMMUNITY PARTICIPATION IN URBAN PEST MANAGEMENT

The need to convey information—knowledge and techniques—from those who have it to those who need it is a common problem in almost every field. To be useful, information must be conveyed understandably, and in such a way that it is likely to be used and used effectively. Although public education in urban pest management is clearly needed, the problems of conveying pest management information to urban residents have not been studied sufficiently to permit definitive statements about which methods assure that the message will be received and applied. Of the small number of existing studies that describe how urban pest management information has been conveyed, some dealt with atypical urban and rural settings, some were questionnaire surveys with very low response rates, and others were conducted so long ago that they are of dubious value today. Hence, there is an urgent need for research on how to transfer information on urban pest management effectively.

Previous research on communicating pest management information in rural areas is largely inapplicable to urban pest management. In rural settings, both the problems and the population addressed are generally homogeneous. Rural pest management is part of the task of monoculture of crops, and thus the population to be addressed has a clear occupational interest in obtaining accurate information. The homogeneity of the rural population and the likely similarity of its educational background and life style allow greater assurance of the appropriateness of the form and medium of the message communicated.

Little of this is applicable to heterogeneous urban populations. The nature of urban pest problems in the inner city slum differs from that of the green suburb, and the pest-related health problems of the inner city are likely to require a different method of communication than the largely economic and aesthetic pest problems of the suburbs. Even within the inner city, moreover, the population has many differences in background, education, and life style. A variety of media and messages—many as yet untried—may therefore be necessary to convey useful information on pest control. Methods will also be needed to measure the impact and effectiveness of particular educational efforts so that more effective means can be devised for future use. As indicated in the working paper by Lenneal Henderson (see Appendix B) prepared for this Committee, there is ample material on the subject of methods of communicating information, but the applicability of available techniques to urban pest management has not been demonstrated.

Effective communication of adequate information is particularly significant if community participation is to be realized. Although the population of urban areas, and particularly the inner-city population, is directly affected by governmental decisions relating to environmental quality, these decisions are generally made by technical agencies and policy makers whose primary concern is usually broader than the neighborhood. Housing and transportation issues usually have greater visibility than more general environmental questions, and inner-city residents are therefore more likely to be given the opportunity to participate in their solution. Inner-city residents need not only more adequate information but also greater opportunities for participating in the formulation and implementation of environmental policy. Because urban pest management is integrally related to the way in which people live, the participation of inner-city residents is needed so that their particular life styles can be adequately taken into account in dealing with pest problems, many of which are inseparable from the rest of the problems of inner city decay. The issue of inner-city participation does not divide along racial or ethnic lines, but is largely socioeconomic in character, and may be related to the sense of political and economic powerlessness that is shared by many inner-city residents.

Thus, what appears to be called for is an effort by local and state agencies to involve inner-city residents in environmental policy decisions and in the implementation of environmental programs that directly affect them. Such efforts must recognize that although inner-city residents depend on the metropolitan economy for employment, education, and public services, inner-city communities retain a separate identity that revolves around existing community-based organizations and institutions. The integrity of inner-city neighborhoods has been recognized in a variety of programs, such as the War on Poverty and the Model Cities program under the Cities Demonstration and Metropolitan Development Act of 1966. What is needed are linkages between areawide and inner-city environmental and health issues so as to give recognition to special inner-city needs without disconnecting them from the problems of the larger metropolitan system and its economic, legal, political, and administrative infrastructures.

There are ways of recognizing the special interests of inner-city communities in most city planning processes, including such traditional mechanisms as citizen advisory committees or such novel methods as inner-city environmental councils. Minority environmental committees and demonstration projects funded by federal agencies to address inner-city problems would encourage citizen participation. There is substantial evidence that community participation in the development and implementation of health-oriented programs involving housing conditions and lead poisoning have been very effective. The involvement of inner-city communities in the organization, planning, and implementation of urban pest management programs would not only provide an effective means for the management of health-related pests but would also provide an opportunity for community involvement in urban environmental programs generally.

RESEARCH PRIORITIES

There is a major need to assess how local governments identify, institutionalize, and use public resources to solve urban pest problems. Most of the previous research on public decision-making processes and outcomes has focused only minimally on program and administrative structure or on decision-making processes as they influence program outcomes. More research on local institutional and planning issues would help remedy the current situation, in which pest management is diffused among numerous agencies and therefore lacks central status and focus.

There is also a need to reassess the legislative and administrative framework that so critically influences local pest management programs. New administrative models that consolidate diffused authority, responsibilities, and resources should be investigated. There is a need for new program designs that take into account problem assessment, planning, training of personnel and above all, encourage innovative approaches to urban pest problems. Broader public involvement is also needed, particularly at the neighborhood level and through groups concerned with environmental issues. A related need, public education and information dissemination, is strongly supported.

In addition, a broad-based pest management assistance program for localities should be developed. At the very least, this should consist of planning grants that would be used to establish pest inventories and to identify pest-related problems and mechanisms for their solution. A second program component would be special demonstration projects for localities particularly interested in serving as test sites for innovative programs, whose experience might then be useful to other localities. Another component of the program should be federal urban pest control grants designed to deal with local problems but also flexible enough that funds could be shifted to other pressing environmental problems when pest problems are resolved. Communities seeking grants should be required to demonstrate priority needs and strong citizen involvement.

A small and experimental demonstration grant program is also urged. These grants should be allocated exclusively to community-based organizations and other self-help and neighborhood organizations to develop pest control programs at the neighborhood level, with the help of technical assistance from experts. These grants could be administered through a neighborhood-oriented agency, such as HUD or the Community Services Administration.

Efforts should be made to develop and implement uniform guidelines for all federal activities and grant-in-aid programs for local pest management. Federal agencies responsible for housing and urban development, transportation, commerce, health, education, defense, and other areas would be expected to promulgate such uniform guidelines.

A program of research on urban pest problems, biology, habitat, ecology, behavior, and management and control methods, as well as effectiveness evaluation (including social and economic impact assessment) should be undertaken to assist local problem-solving activities and federal policy formulation. Federal grants to universities could assist in strengthening local interaction with pest control operators, hospitals, health departments, nurserymen, and others.

Finally, there is a need for revision of model statutes and ordinances associated with both regulatory and program-administrative activities at the state and local levels. Such model statutes should reflect more appropriate institutional structures for carrying out pest control activities and emphasize the use of alternative pest management techniques whenever practical.

NOTES

1. 35 Fed. Reg. 15623 (October 6, 1970).

2. Pub. L. No. 92-516, 86 Stat. 973, as last amended by Pub. L. No. 95-396, 92 Stat. 819 (Sept. 30, 1978), 7 U.S.C. Sec. 136 *et seq.* (cited as FIFRA).

3. FIFRA Sec. 2(t), 7 U.S.C. Sec. 136(t).

4. FIFRA Sec. 2(u), 7 U.S.C. Sec. 136(u).

5. FIFRA Sec. 3(c)(5), 7 U.S.C. Sec. 136a(c)(5).

6. FIFRA Sec. 2(bb), 7 U.S.C. Sec. 136(bb).

7. FIFRA Sec. 3(d)(1)(C), 7 U.S.C. Sec. 136a(d)(1)(C).

8. FIFRA Sec. 4(a)(1), 7 U.S.C. Sec. 136b(a)(1).

9. FIFRA Sec. 6(b), 7 U.S.C. Sec. 136d(b).

10. FIFRA Sec. 28, 7 U.S.C. Sec. 136w-3.

11. *See* Resource Conservation and Recovery Act of 1976, Pub. L. No. 94-580, 90 Stat. 2795, Sec. 1002, Congressional Findings, 42 U.S.C. Sec. 6901 (cited as RCRA). For legislative history, and for earlier federal solid-waste legislation, *see* F. Grad, *Treatise on Environmental Law*, Sec. 4.02 [3].

12. Grad, note 11 *supra*, at Sec. 4.02 [1], [2]; S. Savas, *Evaluating the Organization of Service Delivery: Solid Waste Collection and Disposal*, Chapters 10, 14 (National Science Foundation Grant No. SSH 74-02061 A O 1).

13. EPA Task Force on Environmental Problems, *Our Urban Environment and Our Most Endangered People* 27-29 (1971).

14. RCRA Sec. 4005, 42 U.S.C. Sec. 6945.

15. RCRA Sec. 4004, 42 U.S.C. Sec. 6944.

16. RCRA Sec. 1008(a)(1), (2), 42 U.S.C. Sec. 6907(a)(1),(2).

17. H.R. Rep. No. 94-1491, 94th Cong., 2nd Sess. 37 (1976).

18. 43 Fed. Reg. 4950 (Feb. 6, 1978).

19. National Commission on Urban Problems ("Douglas Commission"), *Building the American City* 40-55 (1968); EPA Task Force on the Environmental Problems of the Inner City, *Our Urban Environment and Our Most Endangered People* 7-22 (1971).

20. 42 U.S.C. Sec. 1467(a).

21. 42 U.S.C. Sec. 1468a.

22. 42 U.S.C. Sec. 1451(c).

23. *E.g.* APHA - PHS Recommended Housing Maintenance and Occupancy Ordinance, Sec. 7.02, 7.03, Sec. 3.05, 3.08-3.11; Southern Standard Housing Code, Sec. 3.08, BOCA Basic Housing Code, Sec. H - 336.0, H - 336.1-336.3. For discussion, *see* Douglas Commission Report, note 19 *supra*, at 273-307. *See also*, Grad, *Legal Remedies for Housing Code Violations*, Research Report No. 14 for the National Commission on Urban Problems (1968).

24. 42 U.S.C. Sec. 5304.

25. 42 U.S.C. Sec. 5305(a)(3).

26. *See* 12 U.S.C. Secs. 1715,1715z-1.

27. *See e.g.*, 42 U.S.C. Sec. 1437f.

28. FHA Handbook, Division 2, Sec. 502 to 506-3.1; Division 6, Sec. 606 to 606-4.1.

29. Foregoing based on working paper by Mark Issacs, HUD (*see* Appendix B of this report).

30. *Id.*

31. 42 U.S.C. Sec. 247b(j)(2).

32. 42 U.S.C. Sec. 254c(a)(4).

33. 21 U.S.C. Sec. 342.

34. For an early prosecution which set the pattern, *see* United States v. Dotterweich, 320 U.S. 277 (1943).

35. 21 U.S.C. Secs. 331, 342(a)(2), 346a(a), 346a(b); *see, e.g.,* Environmental Defense Fund v. U.S. Dept. of Health, Education, and Welfare, 428 F.2d 1083 (D.C. Cir. 1970), discussed at Grad, note 11 *supra,* at Sec. 8.02 [4][d].

36. *See* 16 U.S.C. Secs. 581, 581a, 581b.

37. Information supplied by Philip J. Spear, Senior Director, Research, National Pest Control Association, Inc.

38. Cooperative Forestry Assistance Act of 1978, Pub. L. No. 95-313, 92 Stat. 365, 16 U.S.C. Sec. 2101 *et seq.*

39. The Science and Education Administration (SEA) was established by the Secretary of Agriculture on January 24, 1978 (43 Fed. Reg. 3254) reflecting the consolidation of the former Agricultural Research Service, the Cooperative State Research Service, the Extension Service, and the National Agricultural Library. *See U.S. Government Manual 1979-80,* pages 129-132 (Washington, D.C.: U.S. Government Printing Office, 1979).

40. 42 U.S.C. Sec. 1751 *et seq.*

41. 7 U.S.C. Sec. 75a.

42. Established by the Secretary of Agriculture on March 14, 1977, pursuant to authority contained in 5 U.S.C. Sec. 301 and Reorganization Plan 2 of 1953.

43. 21 U.S.C. Secs. 451-470, 601-695.

44. 16 U.S.C. Sec. 1 *et seq.* The National Park Service (NPS) administers approximately 300 units in the national park system, which are in three categories—natural, historic, and recreational. Urban parks of a conservational and recreational nature include the Golden Gate National Recreation Area in San Francisco and the Gateway National Recreation Area in New York. EPA, in its *Environmental News* of October 3, 1979, announced that it had entered into an interagency agreement with the National Park Service whereby EPA will provide funds for NPS to implement a model program for controlling park pests, such as insects and rats, while reducing unnecessary pesticide use.

45. For the broad authorizations to the Bureau of Indian Affairs in the management of facilities used by Indians, *see e.g.,* 25 U.S.C. Sec. 13, Sec. 2005.

46. The *U.S. Government Manual 1979-80,* refers to cooperative fish and wildlife research units, located at 45 universities, and lists animal damage control as "operational measures through cooperative programs to control predator, rodent and bird depredation on crops and livestock; research or nonlateral control methods and predator-prey relationships." At 332-33.

47. For powers of GSA to manage, control, and maintain government buildings, *see e.g.,* 40 U.S.C. Sec.285, Sec. 490.

48. The activities of the Federal Supply Service within GSA are authorized in 40 U.S.C. Sec. 481.

49. *See* 38 U.S.C. Sec. 1810(b)(4) ("No loan may be guaranteed under this section or made under section 1811 of this title unless . . . the nature and condition of the property is such as to be suitable for dwelling purposes").

50. National Environmental Policy Act of 1969, Pub. L. No. 91-190, 83 Stat. 852, as amended by Pub. L. No. 94-83, 89 Stat. 424, 42 U.S.C. Secs. 4321-4369. The powers of the Council on Environmental Quality may be found in 42 U.S.C. Secs. 4341-4347. For discussion, *see* Grad, note 11 *supra,* at Sec. 9.01[3].

51. CEQ was authorized to issue guidelines by Executive Order No. 11514, 35 Fed. Reg. 4247 (March 5, 1970). By amendment of the Executive Order, by Executive Order No.

11991, 42 Fed. Reg. 26967 (May 25, 1979), the Council is not authorized to issue "Regulations." *See* 40 C.F.R. Parts 1500-1508 for comment, *see* Grad, note 11 *supra* at 9.01[3][e].

52. President's Message to the Congress, Environmental Priorities and Programs, 15 Weekly Comp. of Pres. Doc. 1353, 1368 (Aug. 2, 1979).

53. Memorandum from the President, Integrated Pest Management, 15 Weekly Comp. of Pres. Doc. 1383 (Aug. 2, 1979).

54. For an account of this development in air and water pollution, *see* Grad, note 11 *supra*, at Sec. 2.03 [1][a], Sec. 3.03[1].

55. *See* section on Federal Responsibility, EPA, FIFRA in this chapter.

56. *See* section on Federal Responsibility, EPA, RCRA in this chapter. Note, particularly, that the Resource Conservation and Recovery Act largely tracks the Clean Air Act with respect to requiring implementing state plans, RCRA Secs. 4001-4009, 42 U.S.C. Secs. 6941-6949.

57. FIFRA Sec. 24(b), 7 U.S.C. Sec. 136v(b).

58. FIFRA Sec. 24(a), 7 U.S.C. Sec. 136v(a).

59. FIFRA Sec. 24(c), 7 U.S.C. Sec. 136v(c).

60. S. Rep. No. 92-838 Pt. II, 92d Cong., 2d Sess. 47 (1976).

61. FIFRA Sec. 4(a)(1), 7 U.S.C.A. Sec. 136b(a)(1). Regulations were promulgated in October 1974. 39 Fed. Reg. 36446 (Oct. 9, 1974), 40 C.F.R. Part 171 (1979). The statutory provision was amended by Section 9 of the Federal Pesticide Act of 1978, 92 Stat. 827, to authorize the EPA Administrator to conduct a program for the certification of applicators of pesticides in any state for which a plan for applicator certification has not been approved by the Administrator. The program must conform to the requirements imposed upon the states by Section 4(a)(2) of the Act, 7 U.S.C. Sec. 136b(a)(2).

62. FIFRA Sec. 4(a)(2), 7 U.S.C.A. Sec. 136(a)(2). Such guidelines have been set for state plans. 40 Fed. Reg. 11698 (March 12, 1975), 40 C.F.R. Secs. 171.7-171.8 (1979). Under these regulations, the state plans are required to (1) designate the agency to be responsible for administering the plan; (2) contain assurances that the agency has the legal authority and qualified personnel to carry out the plan; (3) assure adequate funding to administer the plan; (4) provide for the requisite reports to the EPA Administrator; and (5) assure that state certification standards conform with Sec. 4(a)(1) of FIFRA. Note that the EPA had proposed to conduct a federal program for the certification of applicators of restricted-use pesticides in those states and on those Indian reservations where no approved certification plan was in effect. 42 Fed. Reg. 61873 (December 7, 1977). Then, on June 8, 1978, the Agency amended its pesticide regulations by adding a section to enable the Agency to conduct a federal program for the certification of applicators of restricted-use pesticides in states and on Indian reservations that do not have a certification plan in effect. 43 Fed. Reg. 24834 (June 8, 1978); amending 40 C.F.R. Part 171.

63. FIFRA Sec. 4(a)(1), 7 U.S.C. Sec. 136b(a)(1).

64. FIFRA Sec. 4(b), 7 U.S.C. Sec. 136b(b).

65. *Id.*

66. Grad, note 11 *supra*, at Sec. 8.03[1].

67. For discussion of the Model Act, *see* Grad, note 11 *supra*, at Sec. 8.03[3].

68. *See, e.g.,* Arizona, Colorado, Connecticut, Georgia, Indiana, Iowa, Michigan, New Hampshire, New Jersey, and Utah.

69. *See, e.g.,* Alabama, California, Colorado, Florida, Georgia, Hawaii, Idaho, Indiana, Kansas, Kentucky, Louisiana, Maine, Maryland, Massachusetts, Michigan, Minnesota, Mississippi (procedures are for regulation of aerial application of chemicals and pesticides), Montana, New Hampshire, New Jersey, New Mexico, New York, North Carolina,

Oklahoma, Oregon, Rhode Island, South Carolina, South Dakota, Tennessee, Vermont, Washington, and Wisconsin.

70. *E.g.*, Alabama, Connecticut, Florida, Georgia, Kansas, Rhode Island, Texas, Vermont, and Washington.

71. *E.g.*, Illinois, Indiana, Maine, Massachusetts, New Hampshire, and New Mexico.

72. FIFRA Sec. 26, 7 U.S.C. Sec. 136w-1.

73. FIFRA Sec. 23, 7 U.S.C. Sec. 136u.

74. FIFRA Sec. 27(a),(b), 7 U.S.C. Sec. 136w-2(a),(b).

75. FIFRA Sec. 27(c), 7 U.S.C. Sec. 136w-2(c).

76. Grad, note 11 *supra*, at Sec. 8.03[2].

77. *E.g.*, Calif. Health & Safety Code, Sec. 1155.7, 2200-2910, Cal. Govt. Code Sec. 25842.5. Florida Stat. Sec. 388.011 *et seq.*; Oregon Rev. Stat. Sec. 452.010 *et seq.*; Wash. Rev. Code Sec. 70.22.010 *et seq.*

78. *E.g.*, Indiana Code Sec. 16-1-7.3-1 *et seq.*(authorizes local vector control programs); Louisiana and Maryland provided information that this was a matter of local control; New Jersey Rev. Stat. Sec. 26:2-86 (pigeon control, state and local); Sec. 26:3-64 *et seq.*, state authorization of local adoption of public health nuisance code; New York Public Health Law Sec. 608 (state aid for local mosquito and vector control programs); Secs. 1303, 1308 (authorization to deal with public health nuisances); Ohio relies on local nuisance abatement authority, State provides advice, also on hygiene of housing; Oregon Rev. Stat. Sec. 452.010 *et seq.* (vector control, limited to flies and mosquitoes, local authorization); Pennsylvania (Seven regional control associations engage in local vector control activities); Pa. Code Tit. 25, Ch. 243 (provides for local health nuisance control); state assists localities in rat control program; R. I. Gen. Laws Sec. 23-45-1 *et seq.* (State Health Dept. responsible for distribution of funds for rat control program to communities); Tennessee—local control, Department of Public Health, Division of Environmental Sanitation provides consultation on vector control. Texas Civ. Stat. Art. 4477-1 (sets minimum standards for sanitation and health protection measures—including rats and ectoparasites); Wisconsin—pest management a matter of local control.

79. *See, e.g.*, New Jersey, New York, Pennsylvania, Texas, note 78 *supra.*

80. This common pattern need not be reflected in state law; it is usually a matter of general or home rule municipal power.

81. *E.g.*, California, Department of Health, Local Environmental Health Programs Section, *Services in a Local Environmental Health and Sanitation Program* 36-40 (1976) shows a wide division of authority, with Control of rats, mosquitoes, flies and other insects, Cal. Health & Safety Code, listed as the responsibility of mosquito abatement districts, vector control districts or counties; pest abatement generally, Cal. Health & Safety Code Sec. 2800 in pest abatement districts; housing, Cal. Health & Safety Code Secs. 17922, 17961, Uniform Housing Code, H 201, local health officer or housing department; Rodent abatement, Cal. Health & Safety Code, Sec. 1800 *et seq.*; State Dept. of Health, County supervisor, local health officer; Vector surveillance, Cal. Health & Safety Code Sec. 2425, State Department of Health and local health jurisdictions. More complete state control is found, for instance, in Hawaii, which maintains a Vector Control Branch in its State Health Department, enforcing Hawaii Rev. Stat. (Health Law) Sec. 322-1 on abatement of public health nuisances, and Public Health Regulations, Part 10 on vector control. Note that *plant* pest control is in the Department of Agriculture. Other state functions may include the regulation of structural pest managers, *e.g.*, Ariz. Rev. State Sec. 32-2301 *et seq.*

82. *See* section on Federal Responsibility, HUD in this chapter.

83. *See* Grad, note 11 *supra*, at Sec. 4.02[2], [a]. *See also* section on Federal Responsibility, EPA, RCRA in this chapter.

84. *Id.* For the variety of state legislation on the subject, *see* Grad, note 11 *supra*, at Sec. 4.02[2].

85. *Id. See also* Grad, note 11 *supra*, Sec. 4.02[1].

86. *See* section on Federal Responsibility, EPA, RCRA in this chapter.

87. For example, Detroit divides the responsibility for rat control between the Environmental Protection and Maintenance Department (enforcement of requirements for storage of refuse, litter control, and rodent extermination services in public alleys and easements) and the Department of Buildings and Safety (enforcement of rat-proofing requirements and elimination of interior rat infestations), but the Health Department crosses all lines of enforcement to handle emergency rodent and other pest problems.

88. Detroit and Oakland.

89. *See e.g.*, Dallas City Code Chapter 27 Sec. 27-12(b), "Minimum Urban Rehabilitation Standards" (refers to "a person licensed under the Texas Structural Pest Control Act."). Milwaukee's city ordinance covering commercial pest control operations refers to FIFRA's definition of restricted-use pesticides.

90. Oakland-Alameda County (California Health & Safety Code Sec. 1800 *et seq.*); Jacksonville (Florida Stat. Ann. Chapters 381 and 388); New Orleans (Louisiana Sanitary Code Chapter XVIII).

91. Primarily rat control ordinances.

92. New York City Health Code Sec. 151.01 *et seq. Cf.* Fulton County Health Regulation No. 16, Sec. 4(A)(2) ("Every occupant of a dwelling or dwelling unit shall maintain in a condition not conducive to rat infestation those parts of the dwelling, dwelling unit and premises thereof that he occupies and controls"); Milwaukee Rat Control Regulations Sec. 80-48(1) ("whenever any person or persons shall be in actual possession of or have charge, care or control of any property . . . such person or persons shall be deemed and taken to be the owner or owners of such property").

93. *See, e.g.*, New Orleans Code Sec. 54-2; Houston Code of Ordinances, Sec. 21-93.

94. *See, e.g.*, Revised Ordinances of Albuquerque Sec. 6-18-2 *et seq.*; Houston Code of Ordinances, Sec. 10-168(d).

95. *See, e.g.*, Erie County Sanitary Code, Art. XIX, Sec. 2(c); Norfolk City Code Sec. 39-3; Milwaukee Rat Control Regulations Sec. 80-48(5); Minneapolis Health and Sanitation Code Sec. 229.80. But cf. Erie County Sanitary Code, Art. XIX, Sec. 3(c) ("Every occupant of a dwelling unit when required to do so by the Commissioner of Health shall provide rodent stoppage within the unit occupied by him.")

96. *See, e.g.*, Revised Ordinances of Albuquerque Sec. 6-18-6(B); Erie County Sanitary Code, Art. XIX, Sec. 2(f); Houston Code of Ordinances Sec. 10-171(b); Fulton County Health Regulation No. 16, Sec. 4(A)(1).

97. *See, e.g.*, Revised Ordinances of Albuquerque Sec. 6-18-6(B); Erie County Sanitary Code, Art. XIX, Sec. 2(b); Fulton County Health Regulation No. 16, Sec. 4(A)(8).

98. *See, e.g.*, Fulton County Health Regulation No. 16, Sec. 5(A) (occupants of all rat-proofed business buildings required to maintain rat-proof condition); Norfolk City Code Sec. 39-5 (occupant of rat-proofed building responsible for maintenance and repair of rat-proofing). But cf. St. Louis Rat Control Ordinance Sec. 527.090 ("owner or agent of any rat-stopped building shall maintain it in a rat-stopped condition").

99. *See, e.g.*, Fulton County Health Regulation No. 16, Sec. 4(A)(2); Norfolk City Code Sec. 39-4(b); St. Louis Rat Control Ordinance Sec. 527.140.

100. *See, e.g.*, Revised Ordinances of Albuquerque Sec. 6-18-6(B); Fulton County Health Regulation No. 16, Sec. 4(A)(8); Houston Code of Ordinances Sec. 10-168(d).

101. *See, e.g.*, Revised Ordinances of Albuquerque Sec. 6-18-6(D); Cleveland Codified Ordinances Sec. 211.03; Minneapolis Health and Sanitation Code Sec. 229.130; New Orleans Code Sec. 54-24.

102. *See, e.g.*, Erie County Sanitary Code, Art. XIX, Sec. 4(d); Fulton County Health Regulation No. 16, Sec. 8(A); Houston Code of Ordinances Sec. 21-95 (misdemeanor). Norfolk City Code Sec. 39-15 (fine). Milwaukee Rat Control Regulations Sec. 80-48(6) (fine or imprisonment or both).

103. *See, e.g.*, El Paso Health and Sanitation Code Sec. 12-61; Erie County Sanitary Code, Art. XIX, Sec. 4(i); Houston Code of Ordinances Sec. 21-97; New Orleans Code Sec. 54-22.

104. One example is California Health & Safety Code Sec. 2800:

Pest . . . includes any plant, animal, insect, fish, or other matter or material, not under human control, which is offensive to the senses or interferes with the comfortable enjoyment of life, or which is detrimental to the agricultural industry of the state, and is not protected under any other provision of law.

Note that this definition makes no reference to human health. Oakland-Alameda County considers any disease vector a pest.

105. *See, e.g.*, Fulton County Health Regulation No. 2; Denver Revised Municipal Code Sec. 760.1 *et seq.*

106. Fulton County Health Regulation No. 2, Sec. 1(A).

107. *See, e.g.*, El Paso Health and Sanitation Code Secs. 12-56 to 12-60; Houston Code of Ordinances Secs. 21-99 to 21-101; New Orleans Code Secs. 54-5 to 54-21.

108. *See, e.g.*, Milwaukee Code Secs. 79-3 and 79-4.

109. *See, e.g.*, Revised Ordinances of Albuquerque Sec. 6-18-2 (definition of "Rodent Proofing"); Erie County Sanitary Code Art. XIX, Sec. 1(f); St. Louis Rat Control Ordinance Sec. 527.010 (2).

REFERENCES

Bergman, E. (1974) Evaluation of Policy Related Research on Development Controls and Housing Costs. Center for Urban and Regional Studies. Chapel Hill, N.C.: University of North Carolina.

Field, C. and S. Rivkin (1975) The Building Code Burden. Lexington, Mass.: Lexington Books.

Grad, F. (1971-1979) Treatise on Environmental Law. New York: Matthew Bender Company.

Haskell, E. and V.S. Price (1973) State Environmental Management; Case Studies of 9 States. New York: Praeger.

Ingram, H. and J.R. McCain (1977) Federal water resources management: The administrative setting. Public Administration Review 37:448-455.

Olkowski, H. and W. Olkowski (1978) Making the Transition to an Urban IPM Program. Presented at the Conference on Pest Control Strategies for the Future, Denver, Colorado, March 1978. Berkeley, Calif.: John Muir Institute.

Seidel, S. (1979) Housing Costs and Government Regulations. New Brunswick, N.J.: Center for Urban Policy Research.

U.S. Congress, House (1968) Building the American City. National Commission on Urban Problems (Douglas Commission). House Document 91-34. 91st Congress, 1st Session.

U.S. Department of Agriculture (1979) Integrated Pest Management Programs for
   the State Cooperative Extension Services. A Report of the Extension Committee
   on Organization and Policy. Washington, D.C.: U.S. Department of Agricul-
   ture.
U.S. Department of Health, Education, and Welfare (1979) Questions and
   Answers About Numbers of Establishments. Washington, D.C.: Food and Drug
   Administration.
U.S. Department of Housing and Urban Development (1977) 1976 Statistical
   Yearbook, Annual Housing Survey. Washington, D.C.: U.S. Department of
   Housing and Urban Development and U.S. Department of Commerce, Bureau
   of the Census.
U.S. General Accounting Office (1975) Need for Regulating the Food Salvage
   Industry to Prevent Sales of Unwholesome and Misbranded Foods to the Public.
   Report to Congress by the Comptroller General of the United States, MWD-75-
   64. Washington, D.C.: U.S. General Accounting Office.
Ventre, F. (1971) Maintaining Technological Currency in the Local Building Code:
   Patterns of Communication and Influence. Urban Data Service Report.
   Washington, D.C.: International City Management Association.

Organizations and Individuals Contacted for Comment

GOVERNMENT AND INTERNATIONAL ORGANIZATIONS

Agriculture Experimental Station, Feed & Fertilizer Control Service, Texas
Armed Forces Pest Management Board*
Baltimore City Health Department
California Department of Food and Agriculture, Agricultural Chemicals and Feed Division*
California Department of Health
Center for Disease Control (HEW/PHS)*
Council on Environmental Quality
Department of Housing and Urban Development*
Detroit Health Department
Environment Canada, Environmental Protection Service
Florida Department of Agriculture and Consumer Services
Georgia Department of Agriculture, Feed, Fertilizer, & Pesticides Division
Georgia Department of Public Health; Radiological Health Service, Division of Environmental Health
Illinois Department of Agriculture
Massachusetts Department of Public Health
Michigan Department of Agriculture
Michigan Department of Public Health*
National Cancer Institute
National Institute of Environmental Health Sciences
National Institutes of Health*
National Research Council, Canada
National Science Foundation
New York City Department of Health*
New York State Department of Public Health
Occupational Safety and Health Administration
Office of Technology Assessment

255

Ontario Ministry of the Environment*
Texas Department of Health
U.S. Department of Agriculture*
U.S. Department of Commerce
U.S. Department of the Interior*
World Health Organization*

ENVIRONMENTAL AND PUBLIC INTEREST ORGANIZATIONS

Buildings Research Institute/NAE
Center for Science in the Public Interest
Conservation Foundation
Environmental Law Institute
Environmental Policy Center
Health Research Group
National Audubon Society
National Wildlife Federation
Natural Resources Defense Council
Public Interest Research Group
Scientists' Institute of Public Information
Sierra Club
The Council on the Environment of New York City
Urban Environment Foundation*

PROFESSIONAL/SCIENTIFIC/TRADE ORGANIZATIONS

American Association for the Advancement of Science
American Chemical Society
American Institute for Biological Sciences
American Medical Association
American Mosquito Control Association
American Public Health Foundation
American Registry of Professional Entomologists*
American Society for Testing and Materials
Association of American Pesticide Control Officials, Inc.
Comprehensive Cancer Center—Howard University-Georgetown University
Council for Agricultural Science and Technology
Ecological Society of America
Entomological Society of America
Federation of American Scientists
National Agricultural Chemicals Association
National Environmental Health Association
National Forest Products Association
National Pest Control Association*
Resources for the Future
Society of Toxicology
Synthetic Organic Chemicals Manufacturers Association

CONSULTANT/BUSINESS/INDUSTRY ORGANIZATIONS

Arthur D. Little, Inc.
Battelle Columbus Laboratories
Fisons, Inc.*
Midwest Research Institute
Power Spray Technology, Inc.*
Stanford Research Institute

INDIVIDUALS

Adamson, Lucille
Adkisson, Perry L.*
Alexis, Marcus
Baroni, Geno C.
Bennett, Gary
Billick, Irwin
Bowerman, Allan
Brown, Freddie Mae
Brown, Leland R.*
Brown, Richard
Bryant, Rudolph*
Burkholder, Wendell E.*
Calabrese, Edward*
Carnow, Bertran W.
Chadzynski, Lawrence
Curran, Anita S.
Daniels, Paul*
Davis, David*
Davis, Morris E.*
Dougherty, Charlene
Epstein, Samuel S.
Erickson, Fred
Evans, Therman
Feubert, John
Ford, Amasa
Freeman, A. Myrick III*

Gallant, Martin
Goldsmith, Frank
Guido, Mariam
Hartung, Rolf
Henderson, Lenneal*
Herrington, Lee
Hinkle, Maureen*
Holman, M. Carl
Howard, Walter and Rex Marsh
Hunter, Gertrude*
Hunter, John M.*
Jackson, Connie
Johnson, Raymond
Johnson, Rebecca
Joseph, James
Kaplin, Marshall
Karch, Nathan*
Kates, Robert*
Ladd, Florence
Lanier, Marshall
Leigh, Wilhemina
Lincoln, Charles*
Mampe, C. Douglas
Mason, Thomas J.
Miller, Winston
Mills, Ed

Moeller, Dade
Needleman, Herbert
Nelson, Norton
Nisbet, Ian C.T.*
Nunn, Robert
Olkowski, William & Helga*
O'Neal, Rodney*
Paulson, Glen
Penn, Leo
Poland, Jack
Preuss, Peter
Provenzano, George
Selikoff, Irvin J.
Shapiro, Maurice A.
Sowell, Thomas
Stockdale, Jerry*
Trevethan, Josephine A.*
Williams, Junius
Wilson, Billy R.*
Wilson, John T.
Wolfe, Barbara
Wood, F. Eugene*
Young, Larry
Zuniga, Karen*

*Responded to Committee's request for public input. Copies of written replies are on file for inspection at the Environmental Studies Board, Commission on Natural Resources, National Academy of Sciences, Washington, D.C.

B

Working Papers Prepared for the Committee on Urban Pest Management

Cochran, Donald G., "Statement on Genetic Manipulation of Urban Pest Species."

DeGroot, Rodney C. and William C. Feist, "Structural Pests Other Than Termites."

Ehler, L.E., "Biological Control."

Elmore, Clyde, "Urban Pest Management-Weed Sciences."

Farace, Richard V., "Information Transfer in Urban Pest Management."

Fish and Wildlife Service, U.S. Department of the Interior. "The Role of the U.S. Fish and Wildlife Service in Urban Pest Management."

Good, Joseph, "Urban Pest Management Programs of the U.S. Department of Agriculture."

Henderson, Lenneal, "Inner City Participation in Environmental Management and Planning."

Isaacs, Mark, "HUD's (Housing and Urban Development) Role in Urban Pest Management."

Kappus, Karl, and Benjamin Keh, "Human Disease Associated with Animal Pests in the United States."

Kasl, Stanislav V., "The Social Impact of Infestation of Dwelling Unit on Urban Residents."

Koehler, Carl S., "Host Plant Resistance."

LeVeen, E. Phillip and Mary L. Flint, "The Economics of Urban Pest Management" (includes a Case Study of Cockroach Control).

Secretariat, Armed Forces Pest Management Board, "Information on DOD (Department of Defense) Urban Pest Management Efforts for the National Research Council Environmental Studies."

Weiner, Sandford L. and Harvey M. Sapolsky, "Rats, Bats and Bureaucrats: Urban Organization and Pest Control."

Worf, Gayle L., "Urban Pest Management: A Special Consideration of the Management of Plant Pathogens in an Urban Environment."

The papers are available in limited quantity from the Environmental Studies Board, Commission on Natural Resources, National Academy of Sciences, Washington, D.C.

Selected Bibliography on Urban Environmental Research and Policy

Aiken, M. and R. Alford (1970) Community structure and innovation: The case of public housing. American Political Science Review 64(3):843-864.

Alonso, W. (1972) A theory of the urban land market. Pages 104-110, Readings in Urban Economics, edited by M. Edel and J. Rothenberg. New York: Macmillan and Company.

Bailey, M. (1959) Note on the economics of residential zoning and urban renewal. Pages 319-325, Urban Analysis, edited by A.N. Page and W. Seyfried. Chicago: Scott Foresman and Company.

Banfield, E.C. and J.Q. Wilson (1963) City Politics. Cambridge, Mass.: Harvard University Press.

Berry, B. and W.L. Garrison (1958) Recent developments in central place theory. Pages 107-120, Papers and Proceedings of the Regional Science Association 4. Philadelphia, Pa.: Regional Science Association.

Chinitz, B. (1964) City and suburb. In City and Suburb: The Economics of Metropolitan Growth, edited by B. Chinitz. Englewood Cliffs, N.J.: Prentice Hall.

Clark, T.N., ed. (1968) Community Structure and Decisionmaking. San Francisco: Chandler.

Duncan, O.D. and A. Reiss, Jr. (1956) Social Characteristics of Urban and Rural Communities. New York: John Wiley & Sons.

Duncan, B., G. Sabagh, and M.D. Van Arsdol, Jr. (1962) Patterns of city growth. American Journal of Sociology 67:418-429.

Edel, M. (1972) Planning, market or warfare? Recent land use conflict in American cities. Pages 134-150, Readings in Urban Economics, edited by M. Edel and J. Rothenberg. New York: Macmillan and Company.

Firey, W. (1947) Land Use in Central Boston. Cambridge, Mass.: Harvard University Press.

Harris, L. and Associates (1978) A Survey of Citizen Views and Concerns About Urban Life: Final Report. Washington, D.C.: U.S. Department of Housing and Urban Development.

Hauser, P.M. (1965) Urbanization: An overview. Pages 1-47, The Study of Urbanization, edited by P. Hauser and L. Schnore. New York: John Wiley & Sons.

Hawley, A. (1971) Urban Society. New York: Ronald Press.

Hoover, E. (1948) The Location of Economic Activities. New York: McGraw Hill.

Hoover, E.M. and R. Vernon (1959) Anatomy of a Metropolis. Cambridge, Mass: Harvard University Press.

Hoyt, H. (1933) One Hundred Years of Land Values in Chicago. Chicago: University of Chicago Press. (Cited in Laurenti, L. (1970) Theories of race and property values. *In* Urban Analysis, edited by A.N. Page and W. Seyfried. Chicago: Scott Foresman and Company.

Laurenti, L.M. (1970) Effects of nonwhite purchases on market prices of residences. Pages 275-285, Urban Analysis, edited by A.N. Page and W. Seyfried. Chicago: Scott Foresman and Company.

Losch, A. (1938) The nature of economic regions. Southern Economics Journal 5(138):71-78.

Mumford, L. (1955) The natural history of urbanization. *In* Man's Role in Changing the Face of the Earth, edited by W.D. Thomas. Chicago: University of Chicago Press.

Park, R. (1952) Human Communities. New York: Free Press.

Parsons, T. (1951) The Social System. New York: Free Press.

Rein, M. (1973) Dilemmas of Social Reform. 2nd ed. Chicago: Aldine.

Seneca, J. and M.K. Taussig (1974) Environmental Economics. Englewood Cliffs, N.J.: Prentice Hall.

Tonnies, F. (1957) Community and Society. New York: Harper Torchbooks.

D

Major Urban Arthropod Pests in the United States

Tables D.1 through D.9 list the major urban arthropod pests or pest groups in each of the nine established regions of the U.S. Public Health Service's Center for Disease Control, as reported by extension service personnel at land grant institutions. (The pest groups are listed alphabetically, and the number of states reporting appears in parentheses.)

TABLE D.1   Major Arthropod Pest Groups of the New England Region[a]

Major Indoor Pests	Major Outdoor Pests
Ants: Carpenter (6); Misc. (2)	Ants: Carpenter (2); Misc. (2); Pavement (1)
Cockroaches: American (1)	Aphids (3)
Brown-banded (3)	Apple maggots (2)
German (6)	Biting flies: Black fly (5)
Fabric pests: Clothes moth (2)	Mosquito (5)
Dermestidae (5)	Tabanid (3)
Fleas (4)	Chinch bugs (2)
Mites: Clover (3)	Cutworms (2)
Nonbiting flies: Cluster fly (4)	Eastern tent caterpillar (2)
House fly (1)	Millipedes (2)
Powderpost beetle (2)	Ticks: American dog (3)
Silverfish (2)	Misc. (1)
Termites: Misc. (2)	Wasps (5)
Subterranean (2)	White grubs (4)

OTHER REPORTED PESTS

Indoor	Outdoor
Elm leaf beetle	Borers: Misc.
Houseplant pests: Mealybug	Birch leaf miner
Rice weevil	Nonbiting flies: House fly
Spiders: Misc.	Mites
Ticks: Brown dog	Plum curculio
Wasps	Slugs
	Sowbugs/pillbugs
	Whiteflies

NOTE. Pest groups are listed alphabetically. Number of states reporting pests is in parentheses.

[a]New England region: Connecticut, Maine, Massachusetts, New Hampshire, Rhode Island, Vermont.

TABLE D.2   Major Arthropod Pest Groups of the Middle Atlantic Region[a]

Major Indoor Pests	Major Outdoor Pests
Ants: Carpenter (1)	Aphids (2)
Misc. (3)	Birch leaf miner (1)
Cockroaches: Brown-banded (1)	Biting flies: Black flies (1)
German (1)	Mosquitoes (1)
Misc. (2)	Borers: Squash vine (1)
Earwigs (2)	Misc. (2)
Fabric pests: Dermestidae (2)	Cankerworms (1)
Fleas (3)	Chinch bugs (1)
Mealybugs (1)	Earwigs (1)
Millipedes (1)	Gypsy moths (1)
Mites: Clover (1)	Insect galls (1)
Spider (1)	Mites: Red (1)
Nonbiting flies: Misc. (1)	Spider (1)
Powderpost beetle (1)	Root maggots (1)
Stored-product pests (3)	Sawflies (1)
Termites: Subterranean (2)	Scales: Misc. (3)
Misc. (1)	Taxus weevil (2)
Wasps (1)	Tent caterpillar (2)
Whiteflies (1)	Wasps (2)
	White grubs (3)

NOTE. Pest groups are listed alphabetically. Number of states reporting pests is in parentheses.

[a]Middle Atlantic region: New Jersey, New York, Pennsylvania.

TABLE D.3   Major Arthropod Pest Groups of the East North Central
Region[a]

Major Indoor Pests	Major Outdoor Pests
Ants: Carpenter (3); Misc. (4)	Aphids (2)
Cockroaches: American (1)	Bees (2)
Brown-banded (1)	Biting flies: Black fly (1)
German (2)	Mosquito (3)
Misc. (3)	Tabanid (1)
Oriental (1)	Borers: Black vine (1)
Fabric pests: Dermestidae (3)	Bronze birch (1)
Fleas (2)	Misc. (1)
Mites: Clover (1); Misc. (1)	Shade tree (1)
Nonbiting flies: Cluster fly (1)	Cucumber beetle (2)
House fly (1)	Eastern tent caterpillar (2)
Misc. (1)	Scales (2)
Silverfish (2)	Sod webworm (2)
Sowbugs (2)	White grubs (2)
Spiders: Misc. (3)	Wasps (4)
Stored-product pests (4)	

OTHER REPORTED PESTS

Indoor	Outdoor
Centipedes	Ants: Misc.
Elm leaf beetle	Bagworms
Millipedes	Bark beetles
Powderpost beetle	Cabbage worm
Termites: Subterranean	Canker worm
Wasps	Flea beetles
Weevils	Gall insects
	Leaf hopper
	Mites
	Plantbugs
	Spiders: Black widow
	Brown recluse
	Spotted sap beetle
	Sowbugs
	Termites: Misc.
	Ticks: American dog

NOTE. Pest groups are listed alphabetically. Number of states reporting pests is in
parentheses.

[a]East North Central region: Indiana, Illinois, Michigan, Ohio, Wisconsin.

TABLE D.4   Major Arthropod Pest Groups of the West North Central Region[a]

Major Indoor Pests	Major Outdoor Pests
Ants: Carpenter (4); Misc. (4)	Aphids (4)
Thief (2)	Biting flies: Black fly (1)
Box elder bug (4)	Mosquito (6)
Cockroaches: Brown-banded (4)	Stable (2)
German (5)	Box elder bug (4)
Oriental (1)	Cankerworms (3)
Misc. (1)	Elm leaf beetle (4)
Crickets (3)	Millipedes (3)
Elm leaf beetle (4)	Slugs (3)
Fabric pests: Dermestidae (3)	Sod webworms (4)
Fleas (5)	
Nonbiting flies: Face fly (1)	
House fly (1)	
Misc. (4)	
Silverfish (3)	
Spiders: Black widow (1)	
Brown recluse (2)	
Misc. (3)	
Wolf spider (2)	
Stored-product pests (7)	
Termites: Subterranean (3)	

OTHER PESTS MENTIONED

Indoors	Outdoors
Centipedes	Ants
Ground beetles	Asiatic oak weevil
Millipedes	Bagworms
Mites: Clover	Borers: Misc.
Rove beetles	Crickets
Strawberry root weevil	Cutworms
Wasps	Defoliating larvae
	Gall insects
	Grasshoppers
	June beetles
	Leafhoppers
	Leafminers
	Mimosa webworm
	Mites: Clover and spider
	Moths
	Nightcrawlers
	Nonbiting flies: House and misc.
	Sowbugs
	Thrips
	Ticks
	Wasps
	White grubs

NOTE. Pest groups are listed alphabetically. Number of states reporting pests is in parentheses.

[a]West North Central region: Iowa, Kansas, Minnesota, Missouri, Nebraska, North Dakota, South Dakota.

TABLE D.5 Major Arthropod Pest Groups of the South Atlantic Region[a]

Major Indoor Pests	Major Outdoor Pests
Ants: Carpenter (3)	Ants: Carpenter (2)
Crazy (1)	Field (1)
Field (1)	Fire (1)
Fire (1)	Pavement (1)
Household (1)	Aphids (5)
Misc. (2)	Biting flies: Misc. (1)
Pharaoh (2)	Mosquitoes (2)
Biting flies: Mosquito (3)	Chinch bugs (3)
Stable (1)	Fall Army worm (3)
Cockroaches: American (2)	Lace bugs (3)
Brown-banded (1)	Mites: Misc. (2)
German (4)	Spider (2)
Misc. (3)	Nonbiting flies: House fly (2)
Smokey brown (1)	Misc. (2)
Fabric pests: Clothes moth (1)	Scales (5)
Dermestidae (3)	Sod webworm (5)
Fleas (7)	White grubs (6)
House plant pests: Mealybug (1)	Whiteflies (3)
Scale (2)	
Nonbiting flies: House fly (2)	
Misc. (3)	
Old house borer (3)	
Powderpost beetle (4)	
Stored-product pests (7)	
Termites: Drywood (1)	
Eastern subterranean (2)	
Misc. (1)	
Subterranean (6)	
Ticks: American dog (1)	
Brown dog (2)	
Misc. (1)	

OTHER REPORTED PESTS

Indoor	Outdoor
Centipedes	Alfalfa weevil
Millipedes	Carpenter bees
Silverfish	European Elm Bark beetle
Spiders: Misc.	Leafminers
Wasps	Mealybugs
Whiteflies	Mexican bean beetle
	Millipedes
	Mites: Spider
	Mole crickets
	Moths
	Sessiid borers
	Sowbugs/pillbugs
	Termites: Misc.
	Subterranean
	Ticks: American dog
	Misc.
	Thrips

NOTE. Pest groups are listed alphabetically. Number of states reporting pests is in parentheses.

[a]South Atlantic region: Delaware, Florida, Georgia, Maryland, North Carolina, South Carolina, Virginia, West Virginia.

TABLE D.6 Major Arthropod Pest Groups of the East South Central Region[a]

Major Indoor Pests	Major Outdoor Pests
Cockroaches: American (1)	Biting flies: Mosquito (3)
Brown-banded (1)	Borers: Ash-Lilac (1)
German (1)	Dogwood (1)
Oriental (1)	Flat-headed apple tree (1)
Misc. (3)	Misc. (1)
Crickets (2)	Peach tree (1)
Fabric pests: Clothes moth (4)	Root collar (1)
Dermestidae (3)	Chiggers (2)
Fleas (4)	Fleas (2)
Powderpost beetle (3)	Mites: Misc. (1)
Silverfish (2)	Spider (1)
Sowbug/pillbug (2)	Nonbiting flies: House fly (1)
Spiders: Black widow (1)	Misc. (1)
Misc. (1)	Scales (4)
Brown recluse (1)	Sod webworm (2)
Stored-product pests (4)	Ticks: American dog (1)
Termites: Eastern subterranean (1)	Misc. (1)
Misc. (3)	White grubs (2)
	Whiteflies (3)

OTHER REPORTED PESTS

Indoor	Outdoor
Ants: Carpenter	Chinchbugs
Earwigs	Bagworms
Mites: Clover	Gall insects
Millipedes	Ground pearls
Nonbiting flies: Cluster	Lacebugs
Wasps	Slugs/Snails
	Thrips

NOTE. Pest groups are listed alphabetically. Number of states reporting pests is in parentheses.

[a]East South Central region: Alabama, Kentucky, Mississippi, Tennessee.

TABLE D.7   Major Arthropod Pest Groups of the West South Central
Region[a]

Major Indoor Pests	Major Outdoor Pests
Ants: Carpenter (1)	Ants: Fire (1)
Misc. (2)	Texas harvester (1)
Pharaoh (1)	Biting flies: Mosquitoes (4)
Cockroaches: German (2)	Caterpillars: Misc. (2)
Misc. (2)	Fleas (4)
Smokey brown (1)	Mites: Misc. (2)
Fabric pests: Clothes moth (2)	Nonbiting flies: House fly (1)
Dermestidae(2)	Misc. (2)
Fleas (3)	Scales (2)
Misc. (1)	Termites: Misc. (1)
Houseplant pests: Mealybug (2)	Subterranean (1)
Scale (2)	Ticks: American dog (1)
Mites: Clover (1)	Lone star (1)
House dust (1)	Misc. (3)
Spiders: Misc. (1)	Wasps (2)
Brown recluse (1)	
Stored-product pests (3)	
Termites: Formosan (1)	
Misc. (2)	
Subterranean (2)	
Ticks: Brown dog (1)	
Misc. (1)	

OTHER REPORTED PESTS

Indoor	Outdoor
Biting flies: Misc.	Ants: Misc.
Crickets	Aphids
Earwigs	Borers: Misc.
Millipedes	Chiggers
Nonbiting flies: Misc.	Cockroaches: Smokey brown
Powderpost beetles	Crickets
	Earwigs
	Gall insects
	Spiders: Misc.

NOTE. Pest groups are listed alphabetically. Number of states reporting pests is in
parentheses.

[a]West South Central region: Arkansas, Louisiana, Oklahoma, Texas.

TABLE D.8   Major Arthropod Pest Groups of the Mountain Region[a]

Major Indoor Pests	Major Outdoor Pests
Ants: Misc. (5)	Ants: Misc. (3)
Cockroaches: Brown-banded (5)	Aphids (7)
German (2)	Biting flies: Mosquitoes (5)
Misc. (1)	Earwigs (4)
Fabric pests: Clothes moth (1)	Elm leaf beetle (5)
Dermestidae (6)	Grasshoppers (4)
Misc. (1)	Mites: Clover (1)
Mites: Clover (2)	Spider (2)
Spider (1)	Scales: Misc. (3)
Nonbiting flies: House fly (1)	Oyster shell (2)
Misc. (2)	Pine needle (2)
Silverfish (5)	Sowbugs/Pillbugs (3)
Spiders: Black widow (3)	
Misc. (4)	
Stored-product pests (7)	
Termites: Misc. (2)	
Subterranean (2)	

OTHER REPORTED PESTS

Indoor	Outdoor
Army cutworms	Borers: Bronze birch
Biting flies: Mosquitoes	Misc.
Box elder bug	Box elder bugs
Bugs: (misc.)	Cabbage maggots
Collembola	Collembola
Earwigs	Cutworms
Elm leaf beetle	Fall webworms
False chinch bug	False chinch bugs
Firewood insects	Fruit pests: Codling moth
Fleas	Leaf roller
Fungus gnats	Galls: Honey locust pod gall midge
Houseplant pests: Mealybugs	Misc.
Misc.	Leafhoppers
Moths	Lilac leaf miners
Powderpost beetle	Mealybugs
Scorpions	Millipedes
Sowbugs	Mountain pine beetle
Strawberry root weevil	Nightcrawlers
Ticks: Brown dog	Nonbiting flies: House fly
Weevils: Misc.	Pear slugs
	Pine tip moths
	Reduviids
	Snails and slugs
	Sod webworm
	Spiders: Black widow
	Spruce bud worm
	Termites: Subterranean
	Thrips
	Ticks: Brown dog
	Wasps
	Whiteflies
	White grubs

NOTE. Pest groups are listed alphabetically. Number of states reporting pests is in parentheses.

[a]Mountain region: Arizona, Colorado, Montana, Nevada, New Mexico, Utah, Wyoming.

TABLE D.9   Major Arthropod Pest Groups of the Pacific Region[a]

Major Indoor Pests	Major Outdoor Pests
Ants: Argentine (1)	Ants: Argentine (1)
Carpenter (1)	Harvester (1)
Harvester (1)	Misc. (2)
Misc. (2)	Aphids (3)
Moisture (1)	Bark beetles (1)
Cockroaches: Brown-banded (1)	Biting flies: Misc. (1)
German (1)	Mosquitoes (2)
Misc. (2)	Crickets (1)
Oriental (1)	Cutworms (1)
Crickets (2)	Earwigs (2)
Earwigs (2)	Grasshoppers (1)
Fabric pests: Clothes moth (1)	Leafrollers (1)
Dermestidae (2)	Mites: Spider (1)
Firewood insects (1)	Moths (1)
Fleas (3)	Nonbiting flies: Misc. (1)
House plant pests: Mealybug (1)	Oak moths (1)
Scale (1)	Pitch moths (1)
Whitefly (1)	Root maggots (1)
Mites: Spider (1)	Scales (1)
Moths (1)	Slugs (1)
Nonbiting flies: Cluster fly (1)	Snails (1)
House fly (1)	Ticks: Misc. (1)
Misc. (2)	Wasps (3)
Powderpost beetles (1)	Webworm (1)
Silverfish (1)	Whiteflies (2)
Spiders: Black widow (1)	
Brown recluse (1)	
Misc. (3)	
Stored-product pests (3)	
Termites: Dampwood (1)	
Drywood (1)	
Misc. (1)	
Subterranean (3)	
Wasps (1)	

NOTE. Pest groups are listed alphabetically. Number of states reporting pests is in parentheses.

[a]Pacific region: California, Oregon, Washington.

E

Cities and States That Supplied Information on Legal Aspects of Local Pest Management

CITIES:

Albuquerque	Denver	Louisville	Oklahoma City
Atlanta	Detroit	Memphis	Omaha
Austin	El Paso	Milwaukee	Pittsburgh
Buffalo	Honolulu	Minneapolis	St. Louis
Charlotte	Houston	Nashville	San Antonio
Chicago	Indianapolis	Newark	San Diego
Cincinnati	Jacksonville	New Orleans	San Francisco
Cleveland	Kansas City	New York	Toledo
Columbus	Long Beach	Norfolk	Tucson
Dallas	Los Angeles	Oakland	Washington, D.C.

STATES:

Arizona	Indiana	New Jersey	Pennsylvania
California	Kentucky	New Mexico	Rhode Island
Connecticut	Maryland	New York	Tennessee
Florida	Michigan	Ohio	Texas
Georgia	Minnesota	Oklahoma	Virginia
Hawaii	Nebraska	Oregon	Wisconsin

Cities and Counties That Supplied Information on Administrative and Management Aspects of Local Pest Management

Akron, Ohio
Alameda County, California
Albuquerque, New Mexico
Allegheny County, Pennsylvania
Atlanta, Georgia
Baltimore, Maryland
Baton Rouge, Louisiana
Birmingham, Alabama
Bridgeport, Connecticut
Broward County, Florida
Buffalo, New York
Chattanooga, Tennessee
Chicago, Illinois
Cincinnati, Ohio
Cleveland, Ohio
Dallas, Texas
Des Moines, Iowa
Detroit, Michigan
El Paso, Texas
Erie County, Buffalo, New York
Erie County, Pennsylvania
Flint, Genesee County, Michigan
Fort Wayne, Allen County, Indiana

Guilford County, North Carolina
Houston, Texas
Indianapolis, Indiana
Kansas City, Missouri
Kansas City, Wyandotte County, Kansas
Las Vegas, Nevada
Los Angeles County, California
Louisville, Kentucky
Macon, Bibb County, Georgia
Memphis, Shelby County, Tennessee
Milwaukee, Wisconsin
Minneapolis, Minnesota
Mobile County, Alabama
Nashville, Davidson County, Tennessee
New Orleans, Louisiana
New York, New York
Norton, Massachusetts
Omaha, Douglas County, Nebraska
Oneida, New York
Orange County, California
Peoria, Illinois
Philadelphia, Pennsylvania
Portsmouth, Virginia

Some localities reported pest management services by more than one department or agency.

San Francisco, California
Savannah, Chatham County, Georgia
Scranton, Lackawanna County, Pennsylvania
Seattle, King County, Washington
St. Louis County, Minnesota
St. Louis, Missouri
Virginia Beach, Virginia
Washington, D.C.
West Palm Beach, Florida
Wichita, Sedgwick County, Kansas